# NORTH AMERICA AND THE GREAT ICE AGE

**McGRAW-HILL EARTH SCIENCE PAPERBACK SERIES**

Richard Ojakangas, Consulting Editor

Bird and Goodman:   PLATE TECTONICS

Cowen:   HISTORY OF LIFE

Kesler:   OUR FINITE MINERAL RESOURCES

Matsch:   NORTH AMERICA AND THE GREAT ICE AGE

Oakeshott:   VOLCANOES AND EARTHQUAKES:
GEOLOGIC VIOLENCE

Ojakangas and Darby:   THE EARTH: PAST AND PRESENT

# NORTH AMERICA AND THE GREAT ICE AGE

CHARLES L. MATSCH

Geology Department
University of Minnesota at Duluth

## McGRAW-HILL BOOK COMPANY

New York  St. Louis  San Francisco  Auckland  Düsseldorf  Johannesburg
Kuala Lumpur  London  Mexico  Montreal  New Delhi  Panama
Paris  São Paulo  Singapore  Sydney  Tokyo  Toronto

This book was set in Helvetica by Black Dot, Inc.
The editors were Robert H. Summersgill and Carol First;
the designer was J. E. O'Connor;
the production supervisor was Judi Allen.
The drawings were done by Christabel Grant.
Kingsport Press, Inc., was printer and binder.

**NORTH AMERICA AND THE GREAT ICE AGE**

234567890KPKP78321098

Library of Congress Cataloging in Publication Data

Matsch, Charles L
    North America and the great Ice Age.

    (McGraw-Hill earth science paperback series)
    Includes index.
    1.  Glacial epoch—North America.  I.  Title.
QE697.M387        551.7′92′097        75-42329
ISBN 0-07-040935-8

For Mildred L. and Louis B. Matsch,
my mother and father;
and all my sisters and brothers

# CONTENTS

# FOUR

# FIVE

# SIX

# SEVEN

# EIGHT

# NINE

# PREFACE

Many people are fascinated by glaciers and the history of the Ice Age. Unfortunately, most of the information regarding our knowledge of glacier activity and Ice Age events is published in hundreds of scientific articles and dozens of special books written mainly for scientists. Just one comprehensive textbook in North America is available on the subject, and that is most suitable for advanced students. Introductory geology textbooks generally cover the entire subject in a brief chapter or two.

This book is written as an Ice Age primer, designed to present the major principles of geology upon which our knowledge of the Ice Age is based along with a summary of the important events in the geological history of the past several million years. The reader should gain an insight and appreciation for *how* we know what we know and an understanding of the relationships between earth systems.

As an introductory text, the book is written in a fairly nontechnical style for the reader without a science background. It is a short, but comprehensive summary, suitable for use in a special-topics class in college earth science and geology, both for liberal-education students and for majors. I have used the material in the book as the basis for a one-credit course on the Ice Age in a series entitled "Topics in Geology," with good results, judging by student evaluations.

The book is most importantly intended to be useful to everyone interested in natural science and the history of the earth. For all, I hope this summary of the Ice Age will bring an added appreciation for and sensitivity to our natural world.

## ACKNOWLEDGMENTS
I wish to acknowledge with gratitude and great admiration the contributions of the hundreds of individual natural scientists whose discoveries and ideas are the core of this book. Individuals who have given great support to the writing and editing are Richard W. Ojakangas, Robert L. Heller, Barry S. Haskell, and Noel Potter, Jr. Persons who contributed photographs are acknowledged individually in the text. Christabel D. Grant designed and drafted the illustrations with finesse and patience. Finally, my thanks to all the friends whose gently nagging, "How's the book coming along?" eventually inspired me to finish it.

*Charles L. Matsch*

# ONE

## THE HISTORY OF A PREPOSTEROUS IDEA

Twenty thousand years ago the landscape of North America was so different that time travelers from the present would find it very difficult to orient themselves. The Southwestern United States, rather than being a desert, was a land of large freshwater lakes and lush vegetation. In place of the present vast open prairie of the Great Plains there was a dense forest dominated by spruce and birch trees and populated by browsing mammoths, musk-ox, and elk. Even the shape of the continent was different, because a lower sea level resulted in the exposure of thousands of square kilometers of additional land. The most profound differences from the present would be found in the northern part of the continent. There, bordered by a belt of frozen tundra, was an ice sheet that covered more than a third of North America. The entire earth, in fact, was still in the grip of a great Ice Age.

How do we know about the environments and events that occurred so far in the past history of the earth that no observer made a record of them? We have learned to interpret the natural record left behind in such things as sediments, fossils, and landforms. Waves from those ancient lakes in the west eroded the rocky slopes that enclosed them. Sand and silt settled in layers on the lake bottoms, burying the remains of aquatic plants and animals that had lived there. Excavations in old bogs and lakes in the Great Plains reveal the bones of elephants and other animals, along with wood fragments and pollen grains from the forests in which they lived. Fishermen trawling along the Eastern seaboard often snag the rotted stumps of trees many fathoms beneath the sea, certainly evidence of former dry land there. Most impressive of all is the record of the great glaciers themselves. They left a blanket of bouldery sediments, and an ensemble of distinctive landforms that dominate the present surface environment over vast areas. But these same features, which we regard today as such a remarkable heritage of an Ice Age, inspired quite a diversity of explanations during the eighteenth and nineteenth centuries, when geology was a developing science. Tracing the history of the concept of a past Ice Age affords some valuable insights into the methods used by geologists to interpret the record of the rocks.

More than anything else, it was the curious presence of the boulders that triggered the intellectual curiosity which would eventually lead to the recognition of an Ice Age. These rock fragments, some of great size, littered much of the landscape of the Northern Hemisphere. Because so many of them were composed of rock types different from the local bedrock, the boulders earned the name *erratic blocks* (Fig. 1-1). Their obvious journey from faraway places was recognized; however, identifying the transit system by which they had traveled led to a great and long controversy. Just as the

**Figure 1-1** This huge boulder of granite gneiss in Yellowstone National Park was carried at least 25 km by glacier ice. Glacial erratics such as this are common features of ice-sculpted terranes, and their abundant distribution on the landscape of northern Europe resulted in the first recognition of the Great Ice Age. *(Photo by David Kelso.)*

**Figure 1-2** The Deluge, vividly depicted here in a painting by Nicolas Poussin, was the geological agent called upon by diluvialists to explain the erratics and deposits of sediments later to be attributed to glaciers.

causes for the extinction of the dinosaurs have attracted a free-for-all of ideas and opinions, so did the erratics inspire and stimulate imaginative response.

The great size and enormous quantity of erratic boulders, along with the deposits of finer sediments generally associated with them, were so impressive that the earliest explanations for their distribution centered around some kind of catastrophe in the earth's history. One such event, the universal flood so vividly recorded in the Old Testament, was thought by some to have occurred in 2348 B.C. It was logical to suppose that such a great hydrological event would have left some kind of geological record. Thus, the erratics and deposits of stony debris that blanketed the more ancient rocks were attributed to the strong currents of the Deluge (Fig. 1-2). This *diluvial theory* was widely held in both Europe and North America, and even today it has its proponents. Within the framework of this theory, the deposits of rock debris were called *diluvium*, and geologists proceeded to document the patterns of various currents to explain their distribution.

In many places the evidence for a Flood was impressive. Besides the erratics, there were wide valleys in which extremely small rivers now flowed. The valleys must have been eroded by much larger streams in the past. The necessary energy could have been contained in the torrents of the receding waters of the Deluge. Conspicuous on the landscape high above the present shorelines of the ocean were wave-cut notches and old beach deposits, which the diluvialists also ascribed to the Flood. Another kind of evidence was sometimes encountered in excavating the diluvium. In some places beneath the layer of bouldery sediment, a zone of compressed plant and animal remains was encountered. This organic litter came to be known as *Noah's barnyard* and interpreted to be relics of the antidiluvial world.

The diluvial theory drew strong support from the Christian Church because it strengthened the view that the Bible, and especially the Book of Genesis, presents a correct chronology for the origin of the earth. Geologists had been promoting a credibility gap with their observations on the nature and rates of geological processes and with their growing awareness of the complex history recorded in the rocks. By 1800 two opposing views of the earth's past were held. The catastrophists believed that the earth had been subjected to a series of swiftly dealt and destructive cataclysms, each followed by a new creation. Another group, influenced especially by a Scottish geologist, James Hutton, held that the earth had evolved through a long history of processes similar to those presently active: processes such as erosion, sedimentation, and volcanism. This uniformitarian way of thinking questioned the scriptural record.

An English geologist, William Buckland, seeing the diluvial theory as a way to reconcile geology and theology, became its champion. In an address at Oxford in 1819 he concluded:[1]

> The grand fact of a universal deluge at no very remote period is proved on the grounds so decisive and incontrovertible, that, had we never heard of such an event from Scripture, or any other authority, Geology of itself must have called in the assistance of some such catastrophe, to explain the phenomena of diluvial action which are unintelligible without recourse to a deluge exerting its ravages at a period not more ancient than that announced in the Book of Genesis.

Although "diluvial action" was widely accepted to explain the erratic boulders, not all were so sure that it was the result of the Noachian Flood.

Even as Buckland was so strongly promoting the diluvial theory, another hypothesis was emerging and gaining adherents. In America, Daniel Drake in 1817 explained the erratics and drift of the Ohio Valley as being from "large fields of ice in a region far beyond the lakes and floated hither by . . . inundations." No doubt, the suggestion that floating ice was a possible mechanism for the transport of rock fragments was inspired by accounts of polar explorers early in the nineteenth century. Icebergs sighted during these expeditions were observed to have embedded in them large quantities of rock debris. Here was a natural process operating at the present time and capable of transporting and eventually depositing, through melt-out, sedimentary debris similar to the erratics and drift. Why could not such a process have acted in the past?

[1]Quoted in R. J. Chorley, A. J. Dunn, and R. P. Beckinsale, 1964, *The History of the Study of Landforms*, Methuen-Wiley, London.

Just as Buckland was at his height of popularity as a defender of the Flood and other catastrophes, a young Scot, Charles Lyell, was completing his study of classics at Oxford. Lyell attended the lectures of Dr. Buckland, and his interest turned to geology. After extensive travels in Europe and exhaustive research into all the geological literature available at the time, he published in 1830 the first volume of a monumental work, *Principles of Geology*. The basis for the entire book was the uniformitarianism of Hutton. In pointing out the weaknesses of the catastrophic theories, Lyell built a strong case for explaining the geological record in terms of the operation of existing natural processes. He could find but one such process in his own observational experience that could explain the erratic boulders. He proposed that North America and Europe had experienced a great marine submergence, resulting from higher sea levels in the past. Changes in the climate resulted in periodic expansion and contraction of polar ice caps. Warm phases, he suggested, promoted the disintegration of ice into icebergs, which then floated across the seas covering the continents (Fig. 1-3). Eventually, the icebergs melted, resulting in a rain of erratics onto the submerged landscape. The resulting deposits came to be known as *drift*, a term still in use as a general name for glacial sediments. This *iceberg theory* gained strong adherents among the uniformitarians. Many geologists, although rejecting the theory as an explanation for the distribution of all of the drift, adopted floating ice as a geologic process to explain at least some of the occurrences of the erratics.

**Figure 1-3**  Large masses of floating ice carrying stony debris, such as these derived from glaciers along the coast of Alaska, became the basis for the iceberg theory to explain the distribution of drift. *(Photo by Ruth A. M. Schmidt.)*

Another kind of ice activity was easily observed in the Swiss Alps. There, many of the valleys in their higher parts enclosed permanent bodies of moving ice in the form of valley glaciers. Inhabitants of the region who lived or worked near the glaciers surely were aware of the relationships between the large deposits of stony debris along the ice margins and the glaciers themselves. These *moraines* were undoubtedly formed by the activity of moving ice.

Long before Lyell had published his iceberg theory, the importance of glaciers as geological agents had been recognized. James Hutton, the great Scottish geologist, in 1795 and John Playfair in 1802 both credited to glaciers the power to move the large erratic boulders scattered throughout the Alps. These conclusions were based on earlier accounts by several Swiss observers. Within the next two decades, a keen interest in glacier studies in Switzerland by Jean De Charpentier and Ignace Venetz brought them to the conclusion that the landscape bore the marks of past glaciations. Both these men were no doubt influenced in their thinking by conversations with a peasant mountaineer, Jean-Pierre Perraudin.

Perraudin had the good sense to intuitively apply one of the most basic principles of geology to his own observations from years of experience in the mountains and valleys of the Alps. In 1818 he wrote the following words:[2]

> Having long ago observed marks or scars occurring on hard rocks which do not weather (these marks always in the direction of the valleys) and of which I did not know the origin, after much thought I finally decided, after going near the glaciers, that they had been made by the pressure or weight of these masses, of which I find traces at least as far as Champsec. This makes me think that glaciers filled in the past the entire Val de Bagnes, and I am ready to demonstrate this fact to incredulous people by the obvious proof of comparing these marks with those uncovered by glaciers at present.

Perraudin's observations are all the more remarkable when one considers that he in all probability did not know that Hutton had proposed the principle of uniformitarianism, which more simply stated says, "The present is the key to the past." In his intuitive application of uniformitarianism to glacial activity, Perraudin, untrained in geology but intelligent and curious, set the stage for the dramatic announcement, almost 20 years later by Louis Agassiz, a young Swiss zoologist, that the entire earth had experienced a great Ice Age.

In the intervening years, Venetz, a civil engineer, extended the observations begun by Perraudin. By 1829 he concluded before a meeting of the Swiss Society of Natural Sciences that all the erratic blocks in Switzerland and northern Europe had been transported by glaciers. In the audience was Jean De Charpentier, who would next carry the baton in the glacial theory relay. Well-educated and a respected scientist, De Charpentier embarked on a close study of glacial activity, as well as the distribution of erratics and of the scratches and grooves on bedrock. He became convinced of the

[2]Louis Agassiz, *Studies on Glaciers,* ed. & tr. by Albert V. Carozzi, 1967, Hafner Publishing Company, Inc., New York, p. xiii.

soundness of the ideas of Perraudin and Venetz regarding the importance of past glacial activity to explain these phenomena. In July of 1834 De Charpentier traveled to Lucerne to deliver a paper supporting Venetz, whose proposal had not been well received in 1829. He later recounted the following incident which occurred during his journey:[3]

> Traveling through the valley of Hasli and Lungern, I met on the Brunig road a woodcutter from Meiringen. We talked and walked together for a while. As I was examining a large boulder of Grimsel granite, lying next to the path, he said: "There are many stones of that kind around here, but they come from far away, from the Grimsel, because they consist of Geisberger (name for granite in Swiss-German) and the mountains of this vicinity are not made of it."
>
> When I asked him how he thought that these stones had reached their location, he answered without hesitation: "The Grimsel glacier transported and deposited them on both sides of the valley, because that glacier extended in the past as far as the town of Bern, indeed water could not have deposited them at such an elevation above the valley bottom, without filling the lakes (those of Brienz and Thun)."
>
> This good old man would never have dreamed that I was carrying in my pocket a manuscript in favor of his hypothesis. He was greatly astonished when he saw how pleased I was by his geological explanation, and when I gave him some money to drink to the memory of the ancient Grimsel glacier and to the preservation of the Brunig boulders.

The publication of De Charpentier's paper a year later attracted wide attention, and both diluvialists and proponents of the iceberg theory stiffened their defenses against this glacial intrusion into their scientific domain.

One perceptive cry from the wilderness, most likely inspired by the published observations of the two Swiss glaciologists, came in 1832 from a German, Professor A. Bernhardi. He published this explanation for the erratic boulders:[4]

> The polar ice once reached as far as the southern limit of the district (in Germany) which is still marked by the erratics. This ice, in the course of thousands of years, shrank to its present proportions, and the deposits of erratics are nothing less than the moraines which the vast sea of ice deposited in its shrinkage and retreat.

Here for the first time was simply and elegantly stated the glacial theory. For some reason Professor Bernhardi's ideas were not widely noticed or taken seriously.

A staunch supporter of Lyell's icebergs, young Louis Agassiz (Fig. 1-4), visited his friend De Charpentier in 1836, and together they examined the evidence that had been cited to support the glacial theory. Agassiz, not yet 30 years old, held the position of Professor of Natural History at Neuchâtel. He

[3]Ibid., p. xv.

[4]Quoted in R. J. Price, 1973, *Glacial and Fluvioglacial Landforms*, Hafner Publishing Company, Inc., New York.

**Figure 1-4** Portrait of Louis Agassiz, ardent proponent of the glacial theory, at the Unteraar Glacier, Switzerland. Oil painting by Alfred Berthoud (1881), in the Library of the University of Neuchâtel. *(Used with permission of The Macmillan Company, from Louis Agassiz, Studies on Glaciers, Albert V. Carozzi, ed. & tr., 1967, Hafner Publishing Company.)*

already enjoyed a wide recognition in both Great Britain and Europe for his work on fossils. He had no experience with glaciers but was immediately convinced of the correctness of Venetz's and De Charpentier's theory. Upon his return to Neuchâtel, Agassiz found the same kinds of evidence for glacial activity in the surrounding region.

On July 24, 1837, Louis Agassiz rose to deliver the presidential address at the opening session of the Swiss Society of Natural Sciences in the town of Neuchâtel. The distinguished audience, expecting to hear a discourse on fossil fishes, was informed that instead they would hear about "glaciers, moraines, and erratic boulders." Agassiz then proceeded to review the observations of Venetz and De Charpentier relating to the glacial features, especially the polished and striated bedrock and moraines found far from the glaciers themselves. He gave his own support to the interpretation that these features were a reliable indication of the larger extent that ice had covered in the past, in the Alps. He offered refutation of currents and icebergs, and then proposed that the earth had experienced a drop in temperature accompanied by "the formation of the huge masses of ice which covered the earth all over the places where erratic boulders are associated with polished rocks." He grandly envisioned this great ice sheet to have extended from the North Pole to the shores of the Mediterranean Sea.

Because of Agassiz's good connections with the scientific community of Europe, his idea of an Ice Age was widely publicized and discussed. In America, Edward Hitchcock was in the process of compiling a lengthy report on the geology of Massachusetts. His chapter on "Diluvium, or Drift," contains a review of hypotheses to explain its origin. After rejecting the Noachian Deluge on scientific grounds, Hitchcock writes:[5]

> A theory has lately been started to explain diluvial phenomena, founded on the action of glaciers in the Alps. Though not originally proposed, it has been chiefly elucidated and defended by the distinguished naturalist, Agassiz, of Neuchâtel in Switzerland. . . . I confess I have as yet seen only a very brief and evidently imperfect development of this theory, in one or two short notices in the scientific journals. But as I understand it, it seems to me inadequate to explain the *tout ensemble* of diluvial action.

He concluded his discussion with: "To my mind, no theory of diluvial action hitherto proposed is so free from objections that I feel satisfied with it." Only 5 years later Hitchcock would become the first American geologist to completely endorse the glacial theory.

During the next several years, after his dramatic announcement at Neuchâtel, Agassiz undertook a close study of glaciers in the Alps. In 1840, he published an impressively illustrated book, *Études sur les Glaciers*, which contains a large amount of detail on the physical aspects of glaciers and their geological activity (Fig. 1-5). For the first time, geologists not familiar with glaciers had an opportunity to learn from Agassiz's fine observations. The last chapter of the book contains a review of the theories put forth to explain the distribution of erratics and polished bedrock. Marshaling all the evidence, he again proposed that during a recent geological epoch, a great ice sheet had covered much of the earth.

The glacial theory, in principle if not in detail, received many adherents, especially owing to the zealous missionary work by Agassiz himself. Few geologists who came to see the evidence in the glaciated valleys with Agassiz as guide were not convinced of the correctness of his conclusions. Buckland came from England in 1838 and returned to Oxford impressed with the theory. In 1840 Agassiz himself traveled to Scotland, where he read a paper explaining all aspects of the glacial theory. As a result of his visit, both Buckland and Lyell came to support the theory. During the fall of that year Agassiz presented his views to the London Geological Society, using examples from Scotland and Ireland to illustrate the activity of ancient glaciers in the British Isles. Both Buckland and Lyell gave papers supporting Agassiz. With such prestigious advocates, the glacial theory was firmly established in Europe.

[5]Edward Hitchcock, 1841, *The Geology of Massachusetts*.

**Figure 1-5**  This etching of the Glacier of Viesch is one of the fine illustrations included in Louis Agassiz's *Études sur les Glaciers*, published in 1840. *(Used with permission of The Macmillan Company, from Louis Agassiz, Studies on Glaciers, Albert V. Carozzi, ed. & tr., 1967, Hafner Publishing Company.)*

In North America, Professor Hitchcock was completely taken with the theory soon after reading *Études sur les Glaciers* and the abstracts of the papers presented to the Geological Society by Agassiz, Buckland, and Lyell. His endorsement of the glacial theory was given in an address to the Association of American Geologists at Philadelphia in April 1841:[6]

> The recent work of Agassiz, entitled *Études sur les Glaciers*, gives a new aspect to the subject (of diluvial phenomena). It is the result of observations made during five summers in the Alps, especially upon the Glaciers, about which so much has been said, but concerning which so little of geological importance has been known. Henceforth, however, glacial action must form an important chapter in geology. While reading this work and the abstracts of some papers by Agassiz, Buckland, and Lyell, on the evidence of ancient glaciers in Scotland and England, I seemed to be acquiring a new *geological sense*; and I look upon our smoothed and striated rocks, our accumulations of gravel, and the "*tout ensemble,*" of diluvial phenomena,

[6]Ibid.

with new eyes. The fact is, that the history of glaciers is the history of diluvial agency in miniature. The object of Agassiz is, first, to describe the miniature, and then to enlarge the picture till it reaches around the globe.

Hitchcock's "Final Report on the Geology of Massachusetts," was already completed and at the printers. In a lengthy postscript entiled "Glacioaqueous Action between the Tertiary and Historic Periods," the perceptive geologist offered a glacial interpretation for many of the "diluvial" phenomena he had earlier found impossible to explain. Thus, the glacial theory gained a strong foothold in America.

Agassiz himself came to America in 1846. He would eventually turn completely to the study of zoology, leaving the final proof of his Ice Age glaciers to others. But what an exquisite feeling of rightness he must have experienced upon landing for the first time in the New World:[7]

When the steamer stopped at Halifax, eager to set foot on the new continent so full of promise to me, I sprang on shore and started at a brisk pace for the heights above the landing. On the first undisturbed ground, after leaving the town, I was met by the familiar signs, the polished surfaces, the furrows, and the striations, so well known in the Old World; and I became convinced of what I had already anticipated as the logical consequence of my previous investigations, that here also this great agent had been at work.

Within two decades the great geological surveys of the United States would begin and Agassiz's vision would result in a wonderful documentation of the activities of the great Ice Age glaciers in North America.

[7]E. Cary Agassiz, 1886, *Louis Agassiz, His Life and Correspondence*, Houghton Mifflin Company, Boston.

# TWO

ICE—A COLD EARTH MATERIAL

The concept of an Ice Age was originally based on evidence that glaciers had once covered much more of the land surface of the earth than they do at the present time. Besides the expansion of enormous glaciers, colder climates promoted the expansion of sea ice and increased the importance of frost action. The geological and environmental effects of the Ice Age thus extended far beyond the physical limits of the glaciers themselves. In order to appreciate the role of ice in shaping the landscape during the Ice Age, it is necessary to study and understand this cold earth material in its natural habitat.

Ice is restricted in its distribution mainly by temperature and the availability of water. It has a simple chemical composition, $H_2O$, and persists where temperatures remain at 0°C or below for considerable periods of time. Ice is called a crystalline solid because the molecules of water which constitute its mass are arranged in an orderly way. Typically the outward appearance of crystalline substances reflects this internal order; snowflakes and frost form flat, lacelike, or box-shaped crystals with six points or sides (Fig. 2-1). It is said that no two snowflakes are alike, and this statement may indeed be true if just the details of external shape are considered. However, every snow crystal displays the same internal ordering of water molecules.

## GLACIERS

In the atmosphere the solid form of water exists mainly as snow. This "chemical sediment" falls from the ocean of air and reaches the earth's

(a)

(b)

**Figure 2-1** (a) A single snow crystal, showing the characteristic six-sided symmetry. *(Photo by Robert M. Carlson.)* (b) Crystals of glacier ice. Even though they display no external symmetry, these ice crystals are internally composed of an orderly arrangement of water molecules. *(Photo by E. R. LaChapelle.)*

surface in many parts of the world. Episodes of precipitation are seasonal over much of the earth, and generally the snows of winter are converted entirely to liquid and vapor during warmer seasons and thereby returned to the earth's water cycle. Certain parts of the earth's surface, however, afford the proper conditions that allow this cold sediment to persist in the solid state for longer periods of time. In these places, snow piles layer upon sedimentary layer. Eventually certain physical and chemical changes take place in response to the loading stress and to temperature changes; the sediment snow is converted into a mosaic of closely interlocking ice crystals. Such masses of ice, as they become thicker, are stressed to the point where they begin to deform under their own weight. The deformation is expressed as movement of the ice, and at the moment of first motion the ice mass becomes a glacier.

Studies of the layers, or strata, of glaciers show the various stages in the production of glacier ice. They reveal that ice formed by this process of transformation from snow is different from many other kinds of ice, such as lake ice or ice in frozen ground. Bulk densities range from 0.05 g/cm³ (grams per cubic centimeter) for new fluffy snow on the glacier surface to 0.91 g/cm³ for pure glacier ice at lower levels, depending upon the amount of air trapped within the mass. A distinctive intermediate stage in the making of glaciers is reached when all the snow crystals have been modified by crushing and regrowth into spherical grains called *firn*. This transformation may take less than 1 year. Old snowbanks display a similar material, called *corn snow*. Closest packing of these spheres of ice results in a mass with density of about 0.55 g/cm³. Beyond this stage, simple compaction ceases, and further density changes are the direct result of recrystallization.

The conversion of firn into glacier ice is a slower process, the rates dependent especially upon temperature. If the temperature is just at the melting point for ice, partial melting and migration of water are rapid. Enhanced mobility of the water molecules speeds up the processes of recrystallization, or refreezing, and under such conditions the terminal stage of solid glacier ice may be reached in just a few decades. The colder the firn mass the slower will be the metamorphism (change of form). On the great ice sheet in Greenland, where mean annual temperatures reach −30°C, the final product, dense glacier ice, is not realized for hundreds of years. Even so, compared with the thousands of years required for the formation of such rock types as shale and sandstone from unconsolidated mud and sand, glacier formation is accomplished, geologically speaking, practically overnight.

**Kinds of Glaciers**   Glaciers have been categorized in a variety of ways, depending upon such things as their dynamic nature, size and shape, and temperature. Because a mass of ice is considered to be a glacier if it is presently moving, or shows evidence of past movement, it may be *active* or *stagnant*. Stagnant glaciers (some use the term *dead* glaciers) are composed of stalled ice that is generally in a state of melting, whereas active glaciers continue to flow. Commonly, the same body of ice shows both conditions in its various parts; stagnant ice is generally confined to the margins.

Size and form result in a classification that ranges from *cirque glaciers*, which are small masses of ice confined to cup-shaped rock basins in mountain terranes, to *ice sheets*, which are enormous glaciers that cover thousands of square kilometers. Cirque glaciers may expand and flood downslope into valleys, taking the form of long, narrow rivers of ice. In this way they become *valley glaciers*. Where valley glaciers merge at the base of mountain ranges, a large apron of ice is formed. This coalescence of ice is called a *piedmont glacier*. During an episode of continued growth and merging, a high-standing landscape might experience a progression from abundant cirque glaciers to the eventual development of a continuous ice sheet (Fig. 2-2).

Glaciologists have discovered in recent years that the temperature of glacier ice is an important control on its activity. Of course, no glacier can be

(a)

**Figure 2-2** (a) Cirque glaciers, these in the Wrangell Mountains, Alaska, are mainly confined to rock basins in mountain terranes. *(Photo by Noel Potter, Jr.)* (b) Valley glaciers, such as the Kaskawulsh Glacier, St. Elias Mountains, Yukon Territory, generally occupy former river valleys. *(Photo by Noel Potter, Jr.)* (c) Valley glaciers merge at the base of mountain ranges to form piedmont glaciers, Malaspina Glacier, Alaska. *(Photo by Austin Post, United States Geological Survey.)* (d) Eventually glacier ice may accumulate to form ice sheets of enormous volume like this great glacier that covers much of Antarctica. *(Photo by Campbell Craddock.)*

(b)

(c)

(d)

warmer than its melting point, which is 0°C at surface pressures. It is very useful to know how close the ice is to its melting point. *Polar glaciers* are those in which temperature is below the melting point. *Temperature glaciers* maintain a temperature just at the melting point. More informally, these types are called "cold" glaciers and "warm" glaciers. In succeeding sections, temperature effects on glacier activity will be pointed out.

**A Delicate Balance**    Many impressive geological features of the earth result from the cumulative effects of small changes over a very long span of time. The existence of glaciers is very simply dependent upon the difference between the volume of the sediment snow that accumulates in a particular place and the volume that is converted to water and vapor and escapes. If the net result of accumulation versus loss is such that some new snow remains each year, then the sum of the net annual increments taken over a period of years may be a glacier.

The term *accumulation* includes all the processes by which snow and solid ice are added to a glacier, such as snowfall, frozen rain, frost from the vapor state, and snow avalanches. The processes by which snow and ice are lost from the glacier are called *ablation*. The boundary on a glacier separating the area of net gain from that of net loss is called the *equilibrium line*. In most glaciers the greatest loss comes from melting. However, wind erosion, evaporation, and avalanches also can result in the loss of mass. Glaciers that terminate in large bodies of water may spawn icebergs by a process called *calving.* Sometimes enormous chunks of glacier ice are set loose in the world's oceans. The tragic meeting between the ocean liner *Titanic* and a large iceberg freed from the Greenland ice sheet resulted in the loss of over 1,500 lives in 1912. These ice "calves" are huge reservoirs of fresh water, however, and exploiting them for water supplies may very well produce a new breed of ocean-going "cowboys" to rope and tow them to the arid regions of the world (Fig. 2-3).

Glaciologists keep track of the general condition of glaciers by measuring the net gains and losses of mass to the system over a period of time. Although this record keeping might at first sound simple, it is extremely difficult to accomplish with any degree of accuracy. The general condition of many glaciers is determined, however, by taking aerial photographs from year to year. Boundaries of the accumulation area can generally be determined because snow, firn, and glacier ice each have distinctive colors on the photographs (Fig. 2-4).

The trace of the boundary of the previous winter's snow that has not been lost to melting is called the annual snow line. Its position is determined by many factors, including total snowfall, temperature, topographic aspect, and day-to-day weather conditions, such as sunshine versus cloudiness. The snow line is an important benchmark to glaciologists because it affords a simple guide to regional climatic trends. The position of the margins of glaciers over a period of time can also be a useful indicator of the mass budget.

Three budgetary conditions are possible in glacier bookkeeping: (1)

**Figure 2-3** Stranded iceberg "calf" broken from the Thompson Glacier in the background, Axel Heiberg Island, Canadian Arctic. Notice that the layering has been rotated during transport. *(Photo courtesy of Judith A. Niemi.)*

**Figure 2-4** Snow line on the Klinaklini Glacier, Coast Ranges of British Columbia, is the boundary between white and gray tone in the upper left. In the same area, sedimentary layers of firn outcrop on the glacier as subparallel patterns of light and dark gray. *(Photo by D. A. Rahm.)*

accumulation is greater than ablation, with the net economy positive; (2) ablation is greater than accumulation, indicating a negative economy; (3) accumulation just equals ablation and the mass balance is said to be in equilibrium. The first condition is characterized by an increasing volume of ice, and this growth is generally accompanied by an expansion of the glacier margins and a thickening of the ice mass. The resulting invasion of the landscape by ice drastically changes the surface environment, and subsequent chapters in this book will detail the results of glaciation as a geological process.

A negative balance in a glacier's budget results in an annual loss of mass. This loss is reflected in the apparent retreat of the ice margins and a general thinning of the entire glacier. Persistent deficits in the mass budget may eventually result in the complete disappearance of glacier ice. Accompanying this deglaciation is the production of large volumes of meltwater, and the dispersal of this water across the landscape results in geological changes as important as glaciation itself.

When the production of glacier ice is just equal to the volume lost to ablation processes, there is no apparent change in the form or thickness of a glacier. The margins appear to be neither advancing nor retreating. The stable configuration of the glacier body through time masks the continuing dynamic nature of the system, however. The glacier retains its form and areal extent through constant movement of ice to the areas of ablation (Fig. 2-5).

**Glacier Flow**  The movement of glaciers was well documented by human experience long before glaciologists began to prove and survey glaciers with precise instruments. Advancing ice margins during the late sixteenth century in certain parts of the Alps dammed side valleys, resulting in the formation of lakes that drowned settlements and caused flooding down-valley when waters burst catastrophically through the ice dams. Ice itself covered several medieval gold mines, forcing their abandonment in the late 1500s. The mines were recovered in the latter part of the nineteenth century as a result of glacier retreat.

The dynamic nature of glaciers seems a contradiction to the observation that ice is a hard, brittle material that shatters easily in response to a swiftly delivered stress, such as the blow of a hammer or ice pick. In response to small forces applied over long periods, ice behaves differently. When a critical thickness is reached, the individual ice crystals are no longer strong enough to hold up under the increasing weight, and slight shifts take place between molecular layers. The net result of such failure by the billions of

**Figure 2-5**  Is this small valley glacier in the St. Elias Mountains, Yukon Territory, advancing, retreating, or in equilibrium? *(Photo by V. N. Rampton.)*

small crystals in a glacier is perceptible movement at the glacier surface. Slow failure of ice crystals internally may also be accompanied by sliding of the entire glacier at the base of the ice mass. Thus, the velocity observed at the glacier's surface is a combination of internal motions and basal sliding.

Measurement of velocities of glaciers is accomplished by setting carefully located stakes on the glacier surface and also by melting boreholes into the ice and inserting pipes that are easily bent. Maximum velocities are generally maintained near the center of the glacier, where frictional drag is at a minimum. Flow rates fluctuate in daily and seasonal cycles; faster rates are observed during warmer periods when the ice temperature is brought closer to the melting point. Heat apparently weakens the internal resistance of crystals, and the presence of meltwater no doubt enhances the sliding capacity by virtue of its lubricating qualities.

Valley glaciers, and especially outlet glaciers of large ice sheets, attain average velocities up to about 200 meters (m) per year. However, flow rates vary widely, even in different parts of an individual glacier. In steep stretches, such as ice falls, a valley glacier may move as much as 2 kilometers (km) per year. Extraordinary velocities of as much as 7 km in a few months have been observed on some glaciers. These *surges,* as they are called, represent unusual events in the regime of glaciers, and their cause is still not well known. The acceleration of these "galloping glaciers" may result from increases in temperature and formation of water at the base that result in instability of the ice masses.

Generally, "cold" glaciers move more slowly than "warm" glaciers. Thus, the ice sheet in Antarctica, which has temperatures throughout much of its mass far below freezing, moves at rates that average less than a few meters per year. In 1968, however, the ice sheet was penetrated to a depth of 3,000 m by drilling, and the base was found to be at the melting point, with water present in abundance. Some glaciologists believe that such a condition, if developed widely throughout that great glacier mass, could result in a sudden rapid flow of ice into the surrounding Antarctic Ocean. The displaced seawater would raise ocean levels dramatically, resulting in the destruction of coastal cities.

Studies on the distribution of movement in glaciers show that definite patterns exist, related to the form of the glacier and to the location of the boundary between accumulation and ablation. Velocities generally decrease with depth, and there is also a slowing down toward the margins. In the zone of accumulation, velocities tend to increase downstream to the equilibrium line, and after that point motion slows progressively to the margin. Flow above the equilibrium line follows a path toward the bottom of the glacier; below the line upward motions carry ice to replace that lost by ablation. A glacier in absolute balance thus shows no change in form, even though a great amount of ice flows through the system (Fig. 2-6).

**Glacier Ice as a Metamorphic Rock**   Rocks are the natural aggregates of chemical compounds that comprise the solid earth. These earth materials are termed igneous rocks if they have solidified from a molten material called magma; sedimentary rocks if they are chemical precipitates from surface

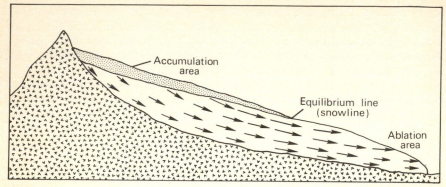

**Figure 2-6** This cross section shows the velocity distribution in an ideal glacier. The length of arrows is proportional to velocity. Movement is downward and accelerating above the equilibrium line, upward and slowing downglacier below that line. In a balanced state, the form of the glacier does not change.

solutions, or cemented fragments of loose debris, such as sand; and metamorphic rocks if the rock has recrystallized from one kind of rock into another. Rocks of the latter origin typically display evidence of having experienced high pressures and temperatures, and some varieties are characterized by features that could have been produced only by internal flow. Glaciers are considered to be metamorphic rocks because they are the result of recrystallization of snow and firn and because they display typical flow structures.

A field description of glacier ice as a rock would include the following observations: (1) It is composed of one mineral—ice—which has the chemical formula $H_2O$. (2) The ice crystals are interlocked into a mosaic texture. (3) Crystal size ranges from a few millimeters up to 0.5 m, with size generally increasing with depth and toward the margins of the ice mass. (4) The ice appears to be banded into discrete layers of alternating coarse-grained, fine-grained, and bubbly ice. (5) The crystals, rather than having a random orientation, have grown in one or several preferred directions.

**Other Features of Glaciers**   Most glacier surfaces exhibit systems of fractures called *crevasses*. A variety of crevasse patterns develops, generally resulting from different flow rates in different parts of the ice. Crevasses reflect the brittle nature of ice; they are rarely more than 30 m in depth because below that thickness, ice deforms by internal motions. In a way these two zones in a glacier, the brittle crevasse-cut outer part and the flowing deeper one, are analogous to the rocks of the earth itself. Our planet has an outer crust, cut by zones of broken rock called faults, and a deeper mantle where rocks deform slowly by flowing.

In steep stretches, especially ice falls, accelerated extending flow produces intensive crevassing and ice thinning. Below such a place, ice slows again and the crevasses close. The opening and closing of crevasses in one season of flow through the ice fall results in the production of a couplet of dark and light ice called an *ogive*. During a summer's flow through the ice fall, the crevasses are open to receive meltwater and rock debris. This ice compresses at the base of the fall to form a dark band. The winter's discharge through the fall is mantled by snow, which fills the crevasses and prevents entry of water and sediment. Compression of this plug of ice produces a light bubbly texture. Ogive patterns are useful in determining rates of flow and distribution of velocity (Fig. 2-7).

## PERMAFROST AND GROUND ICE

Aside from glaciers, cold-climate landscapes are affected by a variety of complicated geological processes that involve ice in other forms. *Permafrost* is a term applied to that condition where ground has remained below 0°C for 2 or more years, regardless of the presence of ice. Where water is present in such areas of cold soil and rock, it crystallizes into *ground ice*. Loose soil thus becomes ice-cemented into solid rock in the same way that sediment in other environments becomes rock by the precipitation of silica or calcite. Presently, vast areas of North America in the arctic regions are underlaid by such frozen soil.

As the condition of permafrost persists, geological processes centered around frost action result in the evolution of distinctive landforms. The growth of large masses of segregated ground ice results in stresses that bulge and crack the surface of the earth, because water expands about 9

**Figure 2-7** Crevasses and ogives, Guyot Glacier, Gulf of Alaska. Ogives are the thick, alternating bands of gray and white. Their trace approximates the velocity profile on the surface, highest in the center. Crevasses demonstrate the brittle nature of the ice. *(Photo by D. A. Rahm.)*

**Figure 2-8** Patterned ground in Alaska. Extreme cold results in contraction of the ground, accompanied by fractures which fill with wedges of ice. Low-centered forms in the upper photo eventually evolve into the high-centered forms in the lower photo as peat accumulates in the center. Diameter of polygons is 15 to 30 m. *(Photo by Ruth A. M. Schmidt.)*

percent when it freezes. That expansion-produced stress is capable of shattering the strongest rocks.

Peculiar patterns of cracks also develop as the ground itself contracts during cooling after it is frozen. These so-called contraction cracks intersect to form a pattern of polygons similar to that formed in the drying sediment of a mud puddle. Eventually these cracks may fill with frozen water to form thick ice wedges (Fig. 2-8).

Thawing of permafrost and ground ice may result in great instability in the capacity of the soil to carry a load. The topography collapses over the melting ice masses, producing a pockmarked terrane. The depressions may become thaw lakes. As ice-cemented sediment thaws, the watery soil may flow slowly down any slope that is present. Imagine the results of thawing of frozen ground upon which roads and buildings have been constructed. Engineers are well aware of the difficulties in constructing facilities in permafrost terranes.

# THREE

## TRACKING ANCIENT GLACIERS

We know that glaciers are capable of causing erosion because the bedrock over which they have moved is left with gouges, scratches, and a fine polish. Valleys through which ice streams flow are deepened and widened into great troughs. Mixed with the ice in glaciers are rock fragments pried from the local bedrock, and the margins of many glaciers are bordered by great piles of rock debris. The latter either fell onto ice streams from the valley sides or were constructed by the glacier conveyor belt as sediment was continuously delivered to a melting terminus. The ice melts and flows away as meltwater, carrying sediments beyond the glacier's margin. As the glaciers themselves melt back, they leave a cover of sediments in their wake.

Thus, the most direct record of past glaciations is found in the erosional and depositional features left behind on the land surface. Mapping the limits of ice-deposited sediments and associated erosional features reveals the maximum extent of Ice Age glaciers. Complex deposits and landforms sometimes contain evidence for more than one advance and retreat of glaciers. Applying some simple geological rules makes it possible to reconstruct the activities of the ancient glaciers that once covered so much of the North American continent. Remember, the present is the key to the past.

### GLACIAL EROSION

Ice is a relatively soft mineral at temperatures near its melting point. Rubbing pure ice against most rocks would produce no visible effects. But *abrasion* by glaciers moving over bedrock becomes effective as an erosional process when stone "tools" are frozen fast into the base of the ice, like the mineral fragments held in sandpaper. Constant movement of the dirty basal ice results in the wearing away of the underlying rock surface. Not all glaciers can abrade. Two conditions must be met: (1) the glacier is sliding along its base; (2) the ice in the basal zone contains rock tools. For sliding to take place, glaciers must be at the melting point. Cold glaciers are frozen onto the bedrock and therefore cannot abrade by sliding. In Greenland, where a tunnel was excavated to the base of a cold ice sheet near its thin margin, the ice was observed to be perched upon lichen-covered bedrock. There was no evidence that the fragile lichens had been disturbed by the great mass of ice moving above the basal zone. In warm glaciers, which are not frozen to bedrock, sliding is enhanced by the lubricating effects of water.

How do rock fragments get mixed into the glacier to provide a potential for abrasion? Although the process is still not completely understood, one possibility is by repeated melting and refreezing of water in the basal zone. For this to occur, the glacier must be very close to the melting point. Slight increases in pressure on the upstream side of bedrock irregularities lower the temperature at which melting occurs. When the pressure is released downstream, the water might refreeze onto the bottom, along with loose rock fragments. The process is similar to that which allows a hockey player to glide across the ice. The pressure of the skate blades lowers the melting point of ice enough to provide a small amount of water, which acts as a lubricant. The water refreezes the instant the pressure is released behind the skater. Once in the mainstream of moving ice, the debris follows flowpaths determined by the overall physics of the glacier, and long-range transport may result (Fig. 3-1).

Abrasion leaves distinctive marks on bedrock surfaces. Parallel scratches, called *striations* (Fig. 3-2), represent the movement of embedded mineral fragments caught between the bedrock surface and the sliding glacier. A single exposure might display hundreds of striations of various lengths, each one determined by the duration of contact of an individual particle. Grooves are features of greater width produced either by tools of greater size or by the enlargement of striations through time. Striated bedrock typically displays polished surfaces, smooth to the touch, as a result of contact with fine-textured abrasives in the ice. Various other marks of small size are found on glaciated bedrock surfaces, in the form of gouges and fractures caused by percussion and sliding of larger rock fragments. These marks are commonly crescent-shaped.

Striations can provide an excellent document of ice-flow patterns. Once the sense of direction of flow is established, mapping the orientation of these marks can lead to the reconstruction of the dispersal patterns of the glacier that made them. Changes in flow patterns through time have been inferred from regional studies that show striations of different orientation crosscutting each other. Successive glaciations have also been interpreted from the presence in a region of crosscutting trends of abrasion marks.

Another common erosional feature of glaciated bedrock terranes is a streamlined form called a *whaleback*, so-called because of its tapered

**Figure 3-1** This 15-m-high ice cliff contains layers of rock debris that have been contorted by glacier flow. In total aspect, the ice-debris mixture is similar to the metamorphic rock called gneiss. Thompson Glacier, Axel Heiberg Island, Canadian Arctic. *(Photo by Judith A. Niemi.)*

**Figure 3-2** Glacial striations and grooves on quartzite bedrock, Mill B Branch, Big Cottonwood Canyon, near Salt Lake City. The bedrock surface has been eroded into streamlined forms called whalebacks, and glacial erratics complete the glacier-produced landscape. *(Photo by R. W. Ojakangas.)*

**Figure 3-3** Large whaleback on Mount Desert Island, Maine. The glacier that shaped this hill of granite bedrock by erosion moved from right to left.

cylindrical shape (Fig. 3-3). Whalebacks are the result of both abrasion and *glacial quarrying*, a process whereby large blocks are plucked from the downglacier end of bedrock hills. The net result is a streamlined, blunt-nosed hill. Because these features record the sense of ice direction (steep end faces downice), they serve a valuable purpose to the glacial geologist. On a practical basis, a knowledgeable outdoorsman can use striations and whalebacks as compasses. For example, in eastern Maine the steep ends of whalebacks invariably face south to southeast and are much more reliable indicators of direction than moss on trees.

## GLACIAL SEDIMENTS

Transportation of sediments by glaciers is analogous in some ways to transportation of sediments by streams. Just like a river, a glacier carries a bedload and a suspended load. Even a small amount of dissolved material is carried in thin films of water separating ice crystals. Eventually rivers deposit their load in a variety of environments, from broad floodplains to deltas, where they enter the sea. Glaciers, too, leave a blanket of bouldery sediments wherever they have flooded across a landscape. Their moraines are arcuate "deltas" marking the position of the glacier margin.

Debris carried by glaciers and deposited directly by the ice is generally a mixture of all sizes of rock fragments, from enormous boulders to fine rock "flour." This sediment, called *till*, is unsorted because ice is so viscous that it is relatively insensitive to differences in weight of the individual fragments (Fig. 3-4a). Glaciers cannot unmix the rock debris, as can running water. Meltwaters produced during ablation, however, selectively winnow out finer sizes and redeposit the sediment in layers beyond the margins of the melting glacier. The result is a sorted deposit called *outwash*, because it literally has been washed out of dirty ice (Fig. 3-4b). Meltwater rivers are generally so charged with the fine rock flour that the water is the color of skim milk (Fig. 3-4c). Seasonal fluctuations of water supply and the abundant load of easily eroded sediments result in the development of a multichanneled, or *braided*,

**Figure 3-4** (*a*) Till deposited at the base of a receding glacier, Axel Heiberg Island, Canadian Arctic. Boulders have diameters of up to 1 m. Meltwater is reworking the till to eventually produce outwash. *(Photo by Judith A. Niemi.)* (*b*) Outwash, sorted by meltwater streams into layers of sand and gravel. (*c*) "Glacial milk," meltwater whitened by suspended rock flour, Athabaska Glacier, Alberta. *(Photo by Noel Potter, Jr.)*

(*a*)

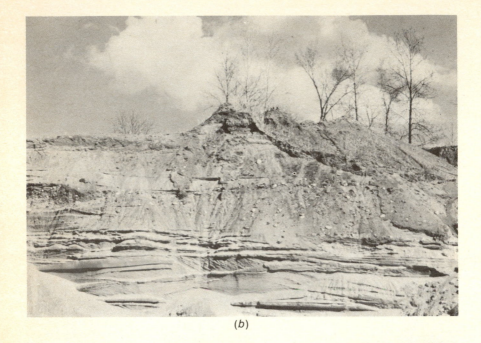

(b)

(c)

**Figure 3-5** Wind blowing fine sand and silt from the outwash deposits along a braided river in the Yukon. *(Photo by V. N. Rampton.)*

stream system. During times of warm and dry weather, wind blowing across the sparsely vegetated deposits of outwash picks up fine sand and silt. The sand might eventually be redeposited to form dunes, and the wind-transported silt accumulates into a blanket of fine-textured loess (Fig. 3-5). If the meltwater enters a lake, the coarser debris is dropped to form a delta, while the finer rock flour settles in deeper water into layers of lake silt. Figure 3-6 shows the general relationship between all of the kinds of glacier-derived sediments.

## TILL

This unsorted and loose rock debris deposited beneath and along the margins of glaciers is the most common deposit in glaciated areas. In gross aspect, the character of till varies considerably from place to place. For example, till in the western Lake Superior region is extremely bouldery and coarse-textured; in contrast, deposits of similar age in eastern North Dakota contain fewer boulders, and the till is composed mostly of silt- and clay-size particles. The difference is explained when the bedrock in each area is examined. The rock formations enclosing Lake Superior are durable igneous and metamorphic varieties containing hard minerals, such as quartz and feldspar. Soft shale and limestone underlie the eastern Dakotas. These rocks have little strength to resist crushing and abrasion. The glacier grinding mill therefore is much more effective in producing a fine-textured product in areas of soft, weak rocks.

**Figure 3-6** The terminus of the Nabesna Glacier, near Northway, Alaska, illustrates the complex sedimentary environments associated with melting glacier ice. Till, outwash, and lake deposits are all intimately associated with various stages of glacier melting. *(Photo by D. A. Rahm.)*

From place to place, therefore, till varies in composition, depending upon the type of rocks encountered by the glacier. The longer the flow path of the glacier, the more likely it is that the till produced will contain a diverse mixture of rock types. Chemical analyses of glacial sediments from Scandinavia were used to make some of the first calculations of the average chemical composition of the earth's crust, on the premise that an ice sheet traveling long distances would fairly sample and mix all the different kinds of earth material it encountered. It has been noted in many places that rocks derived from local bedrock appear in greatest number in the till. Almost every deposit, however, contains some fragments eroded from distant source areas. Agates eroded from the rocks along Lake Superior have been found in glacial sediments near Topeka, Kansas, almost 1,000 km from their source. When the source is known, these *indicators* allow a flow path to be traced. An extensive ice lobe is known to have flowed southeastward from the Dakotas, across Minnesota, and into Iowa because the glacier deposited along its entire path fragments of a particular shale formation restricted to North Dakota. Indicator rock fragments have proved very useful in determining ice flow directions in many parts of North America.

The knowledge that glaciers can sample bedrock by erosion and disperse the eroded material by transport along definable flow paths has been used as an exploration tool for valuable mineral deposits. Astute prospectors have been successful in tracing erratics of such materials back to their bedrock sources, thereby discovering significant ore bodies. One big strike yet to be made is the potentially rich deposit of high-quality diamonds somewhere north of the Great Lakes. For over 100 years, sharp eyes have been spotting diamond fragments in the glacial sediments of the Great Lakes states. Their dispersion pattern points to a source near James Bay (Fig. 3-7).

Several features of till distinguish it from other kinds of unsorted sediment. Individual rock fragments carry facets and scratches resulting from contact with other stones or from being dragged along the bedrock

**Figure 3-7**  Localities where diamonds have been found in glacial sediments. Numbers indicate where more than one have been found. *(Data from C. B. Gunn, 1968, Gems and Gemology.)*

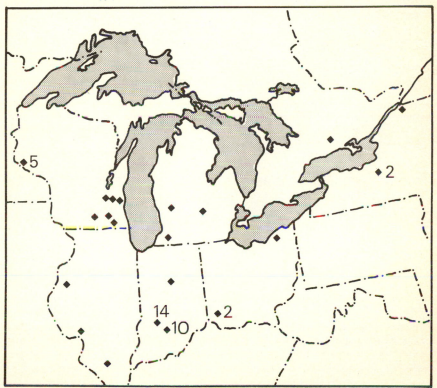

(a)

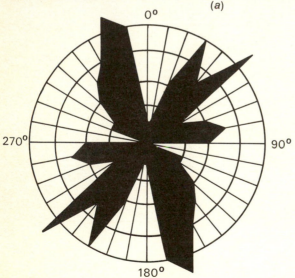

Scale : each circle equals
two stones

(b)

**Figure 3-8** (a) Stones in glacial till at first glance appear to be randomly oriented. Careful measurement of the long axes of elongate rock fragments often reveals a preferred orientation. The rose diagram (b) is a summary of the compass directions of elongated stones, showing concentrations in two general directions.

during transportation. The scratches are not useful as direction indicators because the fragments undoubtedly were rotated about in the glacier. A very interesting aspect of some tills is the alignment of elongate pebbles in several preferred directions. A careful measurement of the position of these rock fragments typically reveals that their long axes have been oriented both parallel to ice flow directions and transverse to the flow direction (Fig. 3-8).

Just why such a pattern exists is not yet understood, but certainly it is a response to the dynamic stresses in the moving glacier ice. Studies on stone orientation in till deposits have been used to help in determining regional flow directions.

## DEPOSITIONAL LANDFORMS

Many northern landscapes owe their striking beauty to the arrangement of sediments left behind during the retreat of glaciers. There is opportunity for glacial till to be deposited at the margins of the ice, to be carried directly to the base of the ice during flow, or to be let down from within or on top of the glacier during ablation. Meltwater has the capability of redistributing sediment in a variety of ways, and no less importantly, gravity induces movement of significant masses of material down the slopes of glacier surfaces. The total result of all these processes is a distinctive suite of landforms after the glaciers have retreated. These constructional features on the earth's surface are valuable indicators of past glacial activity.

*End moraines* are long ridges, or belts of hills, produced by concentrated deposition along the margins of glaciers. Their height, width, and length depend on the size and activity of the glacier which constructed them. The longer the glacial conveyor belt delivers dirty ice to a stabilized margin, the larger will be the debris pile. "Bulldozer-type" action contributes very little to the formation of such features. Receding glacier margins may be marked by successive end moraines, each marking a halt in the general retreat of the ice. When the ice margin retreats uniformly, without intermittent halts, a gently undulating *ground moraine* is left behind. Till associated with ground moraine is generally much thinner than that in end moraine. It is comprised of basal deposits and sediments melted out of the ice during retreat. Both kinds of moraines, especially those associated with continental glaciers, are likely to contain many undrained depressions, resulting from the collapse of sediment over buried stagnant ice blocks that eventually melted. Such *kettle lakes* are distinctive features of glaciated landscapes (Fig. 3-9).

A complicated depositional environment is encountered along the margins of stagnant glaciers. Such a condition generally results in large-scale melting and collapse of debris-laden dead ice, and the simultaneous activity of many depositional processes results eventually in a distinctive *ice-stagnation terrain*. Mudflows slide down wet slopes, meltwater streams wash and sort sediments, and kettle lakes act as sediment traps. Beneath the chaotic surface, differential melting produces a temporary underground system of caverns and tunnels similar to that formed by the solution of limestone in some areas. When all the ice has disintegrated and the collapse is complete, some unusual features result.

Meltwater results in the origin of a variety of landforms. Large, smoothly graded *outwash plains* are formed by deposition from braided streams along the ice margins. Rivers occupying tunnels in the stagnant ice deposit ribbons of sand and gravel along their courses that eventually appear on the landscape as long, sinuous ridges called *eskers* (Fig. 3-10a and b). These distinctive landforms sometimes are many kilometers long, generally not

**Figure 3-9** This landscape of hilly terrain is part of the Alexandria Moraine in west-central Minnesota. The kettle lake is situated in Glacial Lake State Park. *(Photo by David F. Reid.)*

**Figure 3-10** (*a*) Rivers flowing in tunnels within glacier ice deposit gravel and sand melted out of dirty basal ice. Paradise Ice Caves, Mount Rainier, Washington. *(Photo by John Kotar.)* (*b*) Sinuous ridges of sand and gravel, called eskers, are remnants of ice tunnels within glaciers that have since melted away. West-central Minnesota. *(Photo by David F. Reid.)* (*c*) Kames, such as this, are typically composed of sand and gravel washed into depressions on stagnant ice and subsequently unmolded by melting of the enclosing ice walls.

(*a*)

(b)

(c)

more than a few tens of meters wide, and typically as high as they are wide. Eskers contain economically valuable deposits of sand and gravel, and they are so widely exploited in densely populated glaciated areas that as landforms they are in danger of becoming extinct. *Kames* are conical hills of outwash sediment that was originally deposited in openings on the surface of stagnant ice. Melting resulted in their unmolding and eventual addition to the landscape (Fig. 3-10c).

One of the most common landforms associated with glaciation is also the most controversial with regard to its origin. A typical *drumlin* is a streamlined hill, elongate parallel to glacier flow, with a steep slope facing in the upglacier direction. The opposite end displays a gradual taper which points downstream. Drumlins generally occur in swarms numbering into the hundreds. Their extensive distribution in North America has been extremely useful to glacial geologists. However, their origin is still not understood. They are found to consist of till, outwash, or a mixture of these sediment types; and some have a bedrock core. Drumlins are considered to be formed by either subglacial erosion, subglacial deposition, or a combination of both. Whatever their origin, drumlins are highly visible indicators of flow patterns of the ice that formed them (Fig. 3-11).

## EVIDENCE FOR MULTIPLE GLACIATION

Careful observations of glacial sediments and landforms in many parts of North America have resulted in the conclusion that the Ice Age was marked not by just one glacial expansion but by repeated fluctuations of glaciers on the continent. Much of that evidence is in the form of successive layers of till found in some areas, and also in the relationships between suites of landforms.

A very simple geological principle is called the law of superposition. It states that in any undisturbed pile of sedimentary layers, the oldest layer is on the bottom and the strata become progressively younger toward the top. Coupling this means of determining the relative ages of different sedimentary beds with the proper interpretation of their origin allows the determination of a geologic history. Examine the field sketch of an exposure of sediments in southwestern Minnesota shown in Fig. 3-12. A geological interpretation of the sediments indicates four layers of till, each separated by another kind of material. The soils, lake sediments, and outwash all represent times of nonglacial activity, when weathering, erosion, and sedimentation were taking place on an ice-free landscape. Each till marks a glacial advance across the area. Indicator rocks identify the general areas across which the ice advanced, and superposition tells the order in which the events occurred.

A succinct history of the relationship in Fig. 3-12 would go something like this: An advance of ice from the north left a deposit of till. This was followed by glacial retreat. Weathering during a sufficient period of time produced a soil. The soil might even be distinctive enough to indicate the climatic conditions under which it was formed. Another ice lobe advanced from northeastern Minnesota. It may have been a cold glacier because in its movement across the area it did not destroy the soil horizon. This glacier retreated and a lake developed on the deglaciated landscape, followed by a third ice advance, this one again from the north. Meltwaters left a deposit of outwash, which was subsequently overriden by a glacier from eastern North Dakota. Presently the ice-free landscape supports vegetation, and weathering has produced a soil. See how easy it is?

Superposition, although valuable in giving a relative order of events, does not answer other important questions: How much time was involved in

(a)

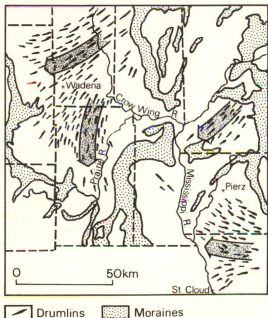

0 _____ 50km

◢ Drumlins    ▦ Moraines

(b)

**Figure** **3-11** (a) Drumlins near Jameson Lake, Washington, show typical streamlined form. Ice movement from bottom to top. *(Photo by D. A. Rahm.)* (b) Regional map of drumlin trends showing general direction of ice movement in central Minnesota. *(From A. F. Schneider, 1961, Minnesota Geological Survey, Bulletin 40.)*

each cycle of activity? Why is there no soil between some of the tills? Is the record here complete, or has erosion removed some of the sedimentary record? Some of these questions can be answered by more careful studies of the sediments across a wider area and by using absolute age-dating

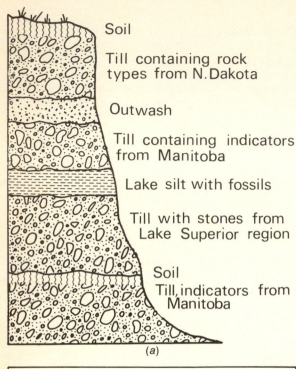

Soil

Till containing rock types from N.Dakota

Outwash

Till containing indicators from Manitoba

Lake silt with fossils

Till with stones from Lake Superior region

Soil

Till, indicators from Manitoba

(a)

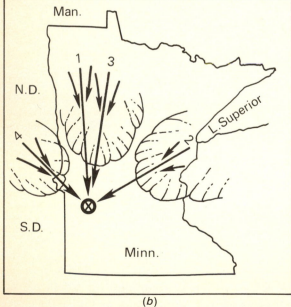

(b)

**Figure 3-12** (a) Field sketch of sediments at location x in southwestern Minnesota. (b) Map shows successive glaciations deduced from the superposition of tills.

techniques (Chap. 4). Other questions might not be answerable at all. Thus, although we can never achieve a complete history of the earth, the broad outlines are generally plain to see.

Another method for determining the relative ages of events is determining crosscutting relationships. If one end moraine trend cuts across another, then it must be younger. Or if a drumlin field trends across an end moraine, the moraine is older. Even with two sets of glacial striations it is possible to determine by this relationship the relative ages of the ice movements that produced them. Combining these principles with an understanding of present-day geological processes produces the key to unlocking the secrets of the earth's history.

The activity of ice sheets in North America was such a recent geological event that many of the resulting landscapes are still relatively unchanged by the postglacial surface modification of weathering and erosion. Because the midcontinent was the southern marginal dumping grounds of the glaciers, very thick deposits of sediments accumulated there, affording an excellent stratigraphic record. Nowhere else in the world is there a better record of continental Ice Age events. The aftermath of those events is a land surface of incomparable variety and a region where glacier-derived resources greatly enhance the quality of human life.

# FOUR

## HOW LONG AGO?
## IN SEARCH OF AN ICE AGE CHRONOLOGY

Before the discovery of radioactivity in 1896, no accurate methods for determining the real ages of rocks were available. Of course, the principles of stratigraphy, and especially observations of crosscutting relationships and superposition in the rock record, had led to the recognition of the relative order of major geologic events. Thus, deposits of the Ice Age, because they are superimposed on all the older rock formations, wherever they are found, must have been the latest sediments to accumulate. The fascinating problem of establishing a chronology in terms of real time provoked a great many imaginative efforts.

**THE AGE OF THE EARTH**
The fact that the earth is at least 4.5 billion years old has only recently been established. James Hutton had concluded in 1788, after recognizing the cyclical nature of rock-forming processes and erosion, that he could find "no vestige of a beginning, no prospect of an end." Others were quite comfortable in the conviction that the earth had been created in the recent past, with no great prospects for a long future existence. The latter attitude was fostered in the conviction that the biblical accounts of the creation were literally correct. A widely accepted birthday for the earth, determined by the

computations based on the genealogies ("begats") in the Book of Genesis by Archbishop Ussher of Ireland, was October 26, 4004 B.C.

Most geologists, who had an opportunity to develop an awareness of the rates of geological processes and the complexity of events recorded in the rocks themselves, rejected such estimates as too short. Such a chronology did not allow enough time for the accumulations of thick sequences of sedimentary rocks, the episodes of folding, metamorphism, and igneous intrusions, and the intervening erosional cycles. Charles Darwin's persuasive arguments supporting the concept of organic evolution appeared in his *Origin of Species*, published in 1859. The acceptance of his theory was contingent upon the availability of an almost limitless amount of time for the random process of natural selection to produce significant changes. The search for a geological timekeeper was vigorously renewed.

To solve the problem, it was necessary to find some natural process which had acted at a constant rate throughout the long history of the earth and which had left a continuous and measurable record. The great thicknesses of sedimentary rocks were first used to solve the problem. By measuring the total thickness of accumulated sediments and determining an average annual rate of deposition, the duration of time represented by the sediments could be deduced. Although this method was simple in concept, its application resulted in a wide range of results because of the errors in measuring sedimentary rock thicknesses and in determining a suitable average rate of accumulation. Certainly no locality on earth has continued to receive sediments during all geological time. Even a casual acquaintance with geological processes imparts an impression of the inconstancy of rates of activity. Thus, it is no wonder that ages for the earth based on rates of accumulation of sediments are not reliable estimates.

Another approach was based on the amount of salt in the ocean. The reasoning behind the method goes like this: Originally the water collected in the ocean basins was fresh, but as weathering and erosion of rocks proceeded, runoff in the form of rivers carried dissolved salts into the ocean reservoir, there to be forever trapped. Measuring the total amount of salt in the ocean today and determining the annual contribution by the world's rivers should allow the calculation of the age of the ocean and, hence, of the earth itself. Allowing for possible recycling of sodium, John Joly published in 1899 his conclusion that the present salt content had taken 90 million years to accumulate. It is not difficult to find fault with his estimate, realizing that ancient seas precipitated enormous quantities of salt desposits which are now part of the sedimentary rocks exposed in continental areas.

The most widely accepted figure for the age of the earth prior to 1900 was that calculated by Lord Kelvin, the great English physicist. It had long been known that the temperature of the earth increases with depth. This *geothermal gradient* means that heat is escaping from the earth's interior to the surface, because heat flows from areas of high temperature to those of lower temperature. Kelvin assumed that the earth had originated as a molten mass which had escaped from the sun. He calculated that 20 to 40 million years would be sufficient time to cool the earth to its present state, assuming that all the heat lost is part of the original energy inherited from the sun. At the time of publication of his final results in 1897, Kelvin did not know that, in the previous year, a discovery had been made in France that would furnish the key to reliable age determinations and refute completely his own work in this area.

In 1896, a French physicist, Henri Becquerel, discovered that compounds of the element uranium produced streaks on photographic plates. The phenomenon was given the name *radioactivity*, and further investigations by a young woman, Marie Curie, and her husband resulted in the discovery of the element radium. The new element generated a measurable amount of heat. In a short time it was realized that radioactive elements in the earth's crust produce enough heat to account for that flowing out at the surface. Lord Kelvin's age estimates were quickly abandoned, and the geologists' concept of a very long time span for the history of the earth regained wide acceptance.

Subsequent discoveries established that a variety of naturally occurring radioactive elements, including isotopes of uranium, thorium, rubidium, and potassium, change as the result of nuclear decay into more stable elements. Rates of radioactive decay have been determined to be constant through time and unaffected by changes in physical and chemical environment. The potential of such material for furnishing the real age of rocks was soon recognized. If the amounts of "parent" material and the "daughter" products that have accumulated in a mineral system as a result of nuclear decay are measured and the rate of radioactive decay is known, then the age of the mineral can be determined. Based on such techniques, thousands of age determinations have been made, and the early ideas of geologists such as Hutton have been confirmed. The earth is billions of years old.

**Subdivisions of Geologic Time**   Long before absolute dating methods were discovered, stratigraphic and fossil studies had resulted in the development of a relative geologic time scale. The largest divisions in this time scale are called *eras*. Shorter *periods* of time were separated on the basis of fossil changes. *Epochs* were proposed as stratigraphic studies showed even more subtle differences in the fossil record. Radioactive age dating allowed the approximate determination of the major boundaries in real years (Fig. 4-1).

The Ice Age in this scheme is formally labeled the Pleistocene Epoch, which is the major division of the Quaternary Period. In another manner of conveying its place in the geologic time scale, the Ice Age transpired late in the Cenozoic Era. Just where to draw the lower boundary for the Pleistocene Epoch in the geologic column has been the focus of disagreement. Traditionally, such boundaries are defined on the basis of changes in the fossil record, and Lyell had originally named and defined the Pleistocene in 1839 on just such a basis. But since that time, most geologists have equated the Pleistocene with ice ages. The difficulty within this context is that climatic cooling is not easy to recognize on a worldwide basis in the sedimentary record, nor is such a condition unique to the Pleistocene Epoch. Such scholarly disagreement acts as a healthy spur to continued investigations into the nature of the earth's past.

| Subdivisions of geologic time | | | Apparent ages (millions of years before the present) |
|---|---|---|---|
| Eras | Periods | Epochs | |
| Cenozoic | Quaternary | Holocene Pleistocene | |
| | Tertiary | Pliocene | 3.0 |
| | | | 13 |
| | | Miocene | |
| | | | 25 |
| | | Oligocene | |
| | | | 35 |
| | | Eocene | |
| | | | 58 |
| | | Paleocene | |
| | | | 65 |
| Mesozoic | Cretaceous | | |
| | | | 135 |
| | Jurassic | | |
| | | | 180 |
| | Triassic | | |
| | | | 225 |
| Paleozoic | Permian | | |
| | | | 280 |
| | Pennsylvanian | | |
| | | | 310 |
| | Mississippian | | |
| | | | 340 |
| | Devonian | | |
| | | | 400 |
| | Silurian | | |
| | | | 430 |
| | Ordovician | | |
| | | | 500 |
| | Cambrian | | |
| | | | 600 |
| Precambrian | | | 4.5 billion |

**Figure 4-1**  Geologic time scale.

## DATING ICE AGE EVENTS

Early attempts at fixing the major events of the Ice Age into a real time frame followed the methods used to determine the age of the earth. Rates of such processes as delta building in lakes left behind by melting glaciers, weathering and soil formation on glacial sediments, peat accumulation in postglacial bogs, and erosion of deglaciated landscapes gave figures that are close to those since determined by radioactive methods. However, all these figures are subject to error because they rely on the assumption that modern rates prevailed in the past.

One very astute attempt to date the close of the Ice Age in North America involved the determination of the retreat of a waterfall on the Mississippi River in Minnesota. The outlet to a large glacial lake, Lake Agassiz, which had formed in the wake of the retreating ice sheet, cut a deep valley across Minnesota. The glacial lake outlet, called River Warren, joined the Mississippi River at St. Paul and eroded a deep channel. The segment of the Mississippi River upstream from its junction with River Warren was undercut, and a waterfall resulted. The waterfall retreated upstream as the bedrock lip eroded under the attack of the turbulent water at the base of the falls, leaving behind an impressive gorge.

An early French missionary-explorer discovered the falls in 1680 and mapped its position, which by then had migrated about 11 km from its point of origin (Fig. 4-2). Subsequent observations fixing its position at various times allowed N. H. Winchell, the first Minnesota state geologist, to calculate a rate of retreat up to the time the falls was stabilized by a dam in 1871. In 1888 Winchell proposed that a total of 7,800 years was represented in the 13-km retreat of the falls. Later corrections to account for changes in bedrock thickness upstream and for a gradual decrease in the height of the

**Figure 4-2** Map showing successive locations during the retreat of St. Anthony Falls. Determining the rate of retreat allowed an estimate of time elapsed since the last glaciation of the Twin Cities. *(Data from N. H. Winchell, 1888, 16th Annual Report, Minnesota Geological and Natural History Survey.)*

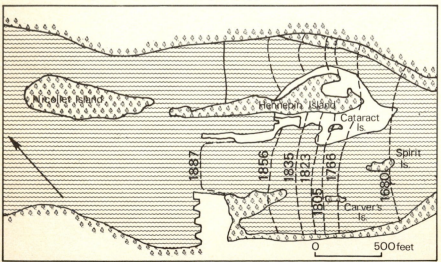

falls resulting from valley filling produced a final estimate of 12,000 years, remarkably close to the presently accepted estimate of 11,700 years.

**Varves**    Thick sequences of sediments in Sweden, consisting of alternating layers of fine clay and coarse silt, have been used to determine the rate of retreat of the Scandinavian ice sheet at the close of the Ice Age. A couplet consisting of one layer each of silt and clay is called a *varve* and is thought to represent the span of 1 year (Fig. 4-3). The origin of varves is linked to seasonal changes in the vicinity of lakes that lie close to glacier margins. During warm periods glacier melting produces runoff charged with rock flour that flows into water ponded near the ice margin. Coarse material settles out first, while finer sediments remain in suspension. Colder temperatures halt the runoff, and the lake may freeze over. During this time, the finer clays settle to form a layer on top of the earlier deposited silt. The spring thaw results in the start of another cycle. Yearly variations in thickness of the varves reflect changes in sediment supply linked to variations in melting of the glacier.

Apparently, the Scandinavian ice sheet left a continuous record of varves in successive lake basins during its general shrinkage. Gerhard de Geer correlated varve sequences from south to north over a distance of almost 1,000 km, matching sequences from deposit to deposit by comparing the pattern of thickness changes in individual varves (Fig. 4-3). As a result of his

**Figure 4-3**    Couplets of dark and light sedimentary layers called varves are thought to represent seasonal changes in a lake environment. If each dark-light band represents 1 year, this sequence of sediments spans about 50 years. St. Croix River Valley near Afton, Minnesota. *(Photo by Junior Hayden.)*

very careful work, he published in 1912 a reliable chronology for the past 12,000 years for Scandinavia, including valuable estimates on the rate of retreat of large ice sheets. His work is confirmed by recent age dating.

A similar study was attempted in North America; unfortunately the development of varves was not so widespread, and the incomplete record, difficult to correlate from place to place. Nevertheless, Ernst Antevs proposed that the last ice sheet to cover North America melted back from a moraine on Long Island to its present position in Greenland during the last 36,000 years. Modern estimates suggest that a period less than half as long was involved.

**Dendrochronology and Lichenometry**   The biosphere affords several methods of limited application for constructing absolute time scales. *Dendrochronology* is the study of tree rings for the purpose of dating past events and reconstructing climatic history. In an annual cycle of growth, a tree under normal conditions adds a layer of large, thin-walled cells formed early in the growing season and finally a layer of smaller, thick-walled cells at the end of the annual cycle. A sharp boundary occurs between the zone of small cells of 1 year's last growth and the large cells of the next year's first growth. The result is a record that is useful in determining the age of the tree (Fig. 4-4). Under certain conditions of stress, trees may not add new growth, and in other instances more than one ring is produced. Therefore, tree-ring analysts exercise great care in establishing their chronologies.

Correlations from tree to tree and from area to area are made on the basis of matching patterns of tree-ring thickness, a method similar to that used in varve studies. This process is best suited to trees whose growth is closely controlled by climatic variability. Where this situation prevails, the width of tree rings is generally a reflection either of the availability of moisture to the tree or of the temperature during the growing season. When responses of individual tree species to different climatic conditions are known, the variations in tree-ring widths can be converted into a continuous report of past climate.

Tree-ring analysis has led to the discovery of the oldest living organisms. Bristlecone pine (*Pinus aristata*) trees older than 4,500 years are growing in parts of the Southwestern United States. Careful correlations between living and dead trees have resulted in a tree-ring chronology spanning the last 7,000 years, and this technique has been used to determine the ages of glacial advances during the last few thousand years.

*Lichenometry* involves dating based on the growth rates of lichens. The time of exposure of rock surfaces to colonization by slow-growing and hardy species, such as *Rhizocarpon geographicum*, can be determined if the growth rate for the lichen is known. Measured are the largest diameters of lichen colonies on rock surfaces of known age, such as mine dumps, quarry walls, or rock cairns (Fig. 4-5). Comparing diameters for lichens growing on moraines or other geological features with the established growth rate curve gives a minimum age. This method is generally not acceptable for dating surfaces more than 3,000 years old. Extrapolations from growth curves of lichens in the Front Range of Colorado suggest the possibility that lichens as old as 6,000 years exist there, rivaling the antiquity of the bristlecone pines.

**Figure 4-4** Prehistoric log section under study in the Laboratory of Tree-Ring Research, University of Arizona. The year A.D. 543 is noted on tree, originally cut in A.D. 623. *(Photo by Bob Broder.)*

**Figure 4-5** The lichen *Rhizocarpon Geographicum*, eastern Baffin Island. *(Photo by G. H. Miller.)*

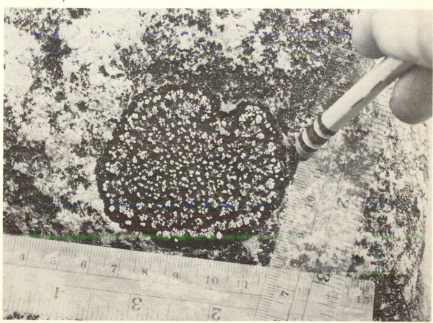

**Radiocarbon Dating**  A radioactive form of carbon called $C^{14}$ (carbon 14) is produced in the upper atmosphere when nitrogen is bombarded by neutrons. $C^{14}$ results when one proton in the nucleus of a nitrogen atom is displaced by a neutron: $N^{14} + n \rightarrow C^{14} + p$. This radioactive carbon rapidly combines with $O_2$ to form carbon dioxide, and becomes part of the atmospheric reservoir. $C^{14}$ atoms are unstable, and they revert back to $N^{14}$ by the emission of electrical energy from the nucleus. In this process of radioactive decay, one neutron is transformed into a proton, and the nucleus becomes a stable nitrogen atom. Careful measurements have determined that one-half of all $C^{14}$ atoms in any sample makes this change in 5,730 years. Knowing this *half-life* allows the age dating of organic remains in the following manner:

Assuming a constant rate of $C^{14}$ production in the atmosphere, the ratio of $C^{14}$ to the much more abundant stable form, $C^{12}$, attains an equilibrium value. Through photosynthesis and the food chain, all living things contain the same proportion of $C^{14}$ to $C^{12}$ atoms. Death halts the continuous exchange, and the radioactive clock is set. As time goes by, $C^{14}$ contained in the dead organism decays at a constant rate, until it has finally all reverted to nitrogen (Fig. 4-6). Measuring the amount of radioactive carbon remaining at any point in time allows the age to be calculated.

Since 1949, when the technique was proposed, thousands of age determinations have been made on such materials as wood, shells ($CaCO_3$), humus in soils, bone material, and even air bubbles trapped in glaciers. Because of the short half-life, even the most sophisticated measuring procedures cannot determine ages beyond about 50,000 years ago. $C^{14}$ chronologies are therefore limited to the last part of the Ice Age. In recent years, variations in the past production of $C^{14}$ have been deduced by dating tree-ring sequences. In this way $C^{14}$ ages were compared with the age of the same wood sample determined by counting tree rings. Ages computed from $C^{14}$ analysis were found to deviate slightly from the tree-ring age. As a result of such studies, correction curves have been constructed from which the $C^{14}$ age can be converted to a real age. Radiocarbon ages greater than 500 B.C.

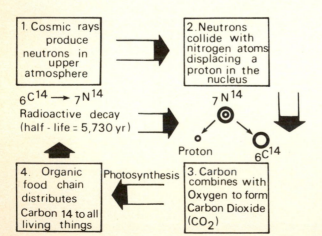

1. Cosmic rays produce neutrons in upper atmosphere

2. Neutrons collide with nitrogen atoms displacing a proton in the nucleus

$_6C^{14} \rightarrow {_7}N^{14}$
Radioactive decay (half-life = 5,730 yr)

$_7N^{14}$

Proton

$_6C^{14}$

4. Organic food chain distributes Carbon 14 to all living things

Photosynthesis

3. Carbon combines with Oxygen to form Carbon Dioxide ($CO_2$)

**Figure 4-6**  Carbon 14 cycle. It is assumed that all living things come into equilibrium with the amount of radioactive carbon in the atmosphere. The total amount of $C^{14}$ is extremely small, less than 0.01 percent of all carbon.

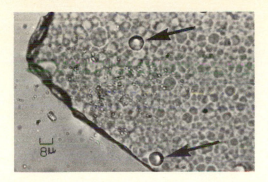

**Figure 4-7** Induced fission tracks in a shard from a volcanic ash deposit. Arrows point to intersected cones etched in the glass by high-energy particles emitted from the nucleus of uranium atoms which have damaged the surrounding mineral matter. *(Photo by John Boellstorff.)*

are generally younger than the real ages. For example, a radiocarbon age of 6,400 years is 7,100 years in real time. Thus, a radiocarbon year does not equal a real year. Such discrepancies may not be important in reconstructing a general geological history, but they must be carefully taken into account in the interpretation of archaic historical events involving human beings.

**Tephrochronology**   Periodic volcanic eruptions during the past have resulted in the deposition of layers of volcanic ash on the earth's surface. Violent explosions witnessed in historical times have distributed such material over large areas many hundreds of kilometers from the volcanic source. Because a single ash layer is the same age every place it occurs, such deposits have great potential value as time markers. They are useful also as a means of determining absolute age because they contain radioactive elements. One technique gaining wide use is called *fission-track dating*. High-energy particles emitted during a spontaneous decay of uranium 238 damage the host mineral crystal along their trajectories (Fig. 4-7). In order to calculate an age for the crystal, the number of spontaneous tracks is counted. Then the uranium concentration is determined by exposing the sample to a neutron bombardment, which induces fission tracks. The age is a function of the ratio of spontaneous tracks to induced tracks.

Two well-known ash falls are used extensively in central and western North America. The Mazama ash covered most of the Northwestern United States and southwestern Canada, following an eruption at Crater Lake, Oregon, about 6,600 radiocarbon years ago. A much older deposit in the central Great Plains, named the Pearlette Ash, has been estimated by fission tracks in mineral grains to be approximately 600,000 years old. This ash originated from volcanoes in the Yellowstone National Park region.

**Other Radioactive Age-Dating Methods**   The use of several other radioactive elements shows promise for determining ages of Ice Age materials. One example involves the decay of potassium 40 ($K^{40}$) into argon 40 ($Ar^{40}$). Because $K^{40}$ is a common constituent of the minerals found in volcanic rocks and it possesses a long half-life, it has become a valuable tool for establishing ages of very old events. The method is hampered because the daughter product of the decay is a gas, and therefore liable to escape from the mineral system. Other techniques involve various daughter products of uranium and thorium found in seawater. These have been used in dating sediments from deep-sea cores.

## PALEOMAGNETIC TIME SCALE

One valuable discovery in recent years is recognition of the fact that the earth's magnetic field has reversed its polarity periodically in the past. Many kinds of rocks and sediments contain a record of the magnetic field extant at the time they were formed. When igneous rocks crystallize from molten magma, certain minerals, such as magnetite ($Fe_3O_4$), tend to align themselves with the prevailing magnetic field, just as do iron filings near a bar magnet. Similarly, fine-grained, iron-rich sediments settling on the ocean floor are influenced in their orientation by magnetic forces. Careful studies of thick volcanic rock and deep-sea sediment sequences, along with absolute age determinations, have resulted in the construction of a geopolarity time sequence. Periods of normal magnetism, times when the poles were similar to the present, alternate with periods of reversed polarity. Matching the magnetic patterns of rocks and sediments from place to place has become an important means of correlating geologic materials of similar age.

## SUMMARY

This chapter can be regarded only as a progress report on the state of a chronology in terms of real time for Ice Age events. Important time gaps remain to be filled with reliable age-dating techniques, and agreement must be reached regarding the stratigraphic definition of the Pleistocene Epoch. Dating deep-sea sediments holds the best promise for establishing a real time scale because climatic fluctuations are recognizable and the sedimentary record is more complete than on the continents (Chap. 5). Continental glaciation probably began as early as 2 million years ago in North America. However, cooler climates are likely to have prevailed even earlier. The latest warming trend is well documented chronologically by $C^{14}$ and other techniques. Important age dates for Ice Age events will be found in succeeding chapters as the geologic history is unfolded.

# FIVE

## EVIDENCE FOR PAST CLIMATIC CHANGE

Nothing is as changeable as the weather, yet the patterns of cold and warm, wet and dry, cloudiness and sunshine, calm and wind have predictable ranges of intensity and reliable intervals of duration and recurrence. Surface processes, landforms, and living communities all are adjusted to the patterns of day-to-day weather that constitute the climate of a region. The study of past climatic conditions indicates that dramatic changes have occurred that had a profound influence on the earth's surface environment. Paleoclimatologists not only attempt to document the nature and magnitude of past climatic shifts but also hope to recognize cyclical patterns that will allow long-range forecasts for the future. The last 2 million years of earth history was a time characterized by climatic changes of planetary extent; the documentation of these changes is a fascinating detective game, with clues hidden in a wonderful variety of natural materials.

### EVIDENCE IN DRIFT

Recognition by early geologists of ancient glacial deposits and landforms far beyond the limits of present-day glaciers led to the conclusion that a colder climate had prevailed in the recent past. Soon after mapping and stratigraphic studies had begun on the glacial deposits, geologists discovered that some sections of drift contained beds of plant debris and soil zones between layers of till. Interpreting the organic and weathered zones correctly as representing

**Figure 5-1** The dark zone is an organic-rich soil horizon developed on glacial drift and buried by till. The soil represents a period of warm climate between glaciations. Southwestern Minnesota.

a nonglacial environment resulted in the recognition that glaciers had advanced and retreated across North America more than once during the Ice Age (Fig. 5-1). The midcontinent region is especially rich in both glacial sediments and landforms, and here a major subdivision of the Pleistocene Epoch evolved.

By the early 1900s four major drift sheets had been distinguished, each separated from the next by some record of a nonglacial environment. These divisions are based on climatic changes from glacial to interglacial conditions and each interval is named for a geographic area:

Youngest

        Present Interglaciation

    Wisconsin Glaciation

        Sangamon Interglaciation

    Illinoian Glaciation

        Yarmouth Interglaciation

    Kansan Glaciation

        Aftonian Interglaciation

Oldest    Nebraskan Glaciation

Studies in Europe confirmed the complex nature of glacial activity there as well. Thus, the stratigraphic record of the glacial sediments themselves contains strong evidence for periodic changes in world climate.

## SEA-LEVEL FLUCTUATIONS

Not long after Louis Agassiz had presented the glacial theory to explain the distribution of erratics, a test of its validity based on studies of sea-level changes was proposed. The reasoning went like this: If ice sheets had covered a much greater surface area in the past, then these ancient glaciers contained enormous amounts of water. Because the water comes from the evaporation of the ocean reservoirs, its storage in land ice represents a temporary loss to the oceans. That loss from the oceans should be reflected in a lower sea level throughout the world. The melting of glacier ice at the end of the Ice Age would result in the return of water to the ocean, resulting in a rise in sea level. In summary, the growth and decay of glaciers on the continents must be accompanied by a fall and rise of sea level.

Many studies since have proved beyond doubt that the level of the sea has indeed fluctuated as predicted. It has been estimated that at the time of the last maximum glacial expansion, water was stored in continental glaciers to the extent that the average level of the sea dropped about 130 m. The most dramatic result of such a change is the exposure of new land around the fringes of continents. In eastern North America the broad continental shelf, which today provides some of the best fishing grounds in the world, was a broad, flat plain, densely forested and populated by a host of land animals. The Ice Age shoreline along the Gulf of Mexico was located hundreds of kilometers seaward from its present position.

Changes in the total volume of water stored in the sea are referred to as *eustatic* changes. Large fluctuations related to the activity of glaciers become a valuable record of worldwide climatic change. However, what at first appears to be a relatively simple tool for documenting past climatic events becomes extremely complicated because of another factor. The best record of past sea levels is found in the erosional and depositional features left as shorelines on the continents. The first reaction to the discovery of a wave-cut feature high above present sea level is to conclude that the mark represents a higher stand of the ocean in the past. But an alternative explanation for its high position is that the land has risen relatively to a stable sea level. Because parts of the crust of the earth are constantly rising and sinking in response to gravity and internal stresses, glacier-controlled sea-level changes are very difficult to determine.

A. L. Bloom, of Cornell University, has found a natural "dip stick" for measuring eustatic changes along parts of the New Guinea coast in the South Pacific. There, mountain-building forces are lifting the land surface at a slow and apparently constant rate. Coral colonies grow in the warm, shallow waters fringing the coastline at the present time, forming a reef complex. Radioactive uranium and thorium, dissolved in minute quantities in sea water, are incorporated into the mineral structure of the reef organisms. In the hills above the coast is a sequence of ancient reef terraces constructed during earlier times and uplifted to their present heights. Reconstructing former sea levels is based on the following reasoning: When sea level is high, signifying melted glaciers and a warm period, a reef complex is built. Cold periods, which are accompanied by glacier growth, result in sea-level lowering and exposure of the reef complex. Uplift slowly raises the coast so that by the time the next high sea level returns, the old terrace is too high to be flooded and a new reef complex is constructed below the old one. Measuring the altitude of the uplifted reefs and age-dating them by the decay

**Figure 5-2** Uplifted reef terraces on the north coast of the Huron Peninsula in New Guinea. Terraces in the lower third of the formation have ages between about 125,000 and 6,000 years. *(Photo by A. L. Bloom.)*

of uranium and thorium allow the estimation of rates of uplift and the level of the sea at the time each reef was formed (Fig. 5-2). This kind of study holds great promise for establishing a chronology for sea-level changes and, by extension, for major cycles of climatic change.

## FOSSIL EVIDENCE ON THE LAND

Vegetation on the surface of the earth is broadly zoned in adjustment to the present climatic belts. Equally adapted are the ranges of many animal species. A change in present climatic patterns would elicit a response from the plant and animal communities in a direction determined by the nature of the change.

An especially useful tool in determining past changes in vegetation in a region is pollen analysis. All flowering plants produce pollen as part of their reproductive process. Transported by the wind, the pollen rains down on the local landscape, most of it missing its intended mark. That which falls onto lakes and bogs quickly becomes incorporated into the mud at the bottom. Because the composition of the pollen rain normally is determined by the kinds and quantities of plants and trees in the vicinity, the deposited sample becomes a record of their distribution. Changes in vegetation are reflected in the pollen production; thus, lake sediments are a valuable storehouse of information to which annual deposits are added.

Probes into the lake-bottom sediments are made to retrieve continuous cores. The thicker the sediments, the longer the time interval spanned, if reasonably constant rates of sedimentation are assumed. Careful identification and counting of the pollen grains from bottom to top allows the monitoring of changes in vegetation by careful interpretation of the results (Fig. 5-3). Skilled pollen analysts are often able to deduce the nature of the climatic changes represented by the change in vegetation inferred from the pollen record. Pollen studies throughout North America indicate that extensive shifts in vegetation zones occurred during the Ice Age in step with the advance and retreat of continental glaciers.

Paleontologists have determined from the study of the fossil remains of many animal species that their ranges have changed in the past. Bones of the musk-ox, now restricted to the tundra of the high Arctic, have been found in Colorado, Iowa, and Pennsylvania. Certain species of lemming, another modern inhabitant of Arctic terranes, are found to have ranged as far south as the central Appalachians. Many other fossil remains suggest repeated migrations of land animals in response to past climatic changes.

**Fossil-Frost Phenomena**   Present-day cold climates are marked by distinctive geologic processes and landforms, already mentioned in Chap. 2. Fossil ice wedges in Wisconsin and Illinois, both localities far south of the present permafrost boundary, indicate a lower mean annual temperature in the past (Fig. 5-4). Remnants of landforms similar to features actively forming today in the Arctic are found in many areas where frost action is not presently an effective geological process. Such features are accurate indicators of cold temperatures, because they form only when the mean annual temperature is below 0°C. Ice-wedge formation today in Alaska is active only in areas where a mean temperature of −6°C is maintained.

## ARIDITY AND ANCIENT WIND DIRECTIONS

Wind as a geologic agent is most important in regions where rainfall is low, evaporation rates are high, and vegetation is sparse. Erosional effects and deposits thus are climatic indicators. In addition, the form of certain sand dunes is related to the prevailing wind direction. Fossil dune fields in western Nebraska indicate a warmer, drier climate in the past. One group of dunes indicates a prevailing wind direction from the northwest (Fig. 5-5), similar to the modern wind pattern. An older group of dunes, however, points to a more northerly prevailing wind direction.

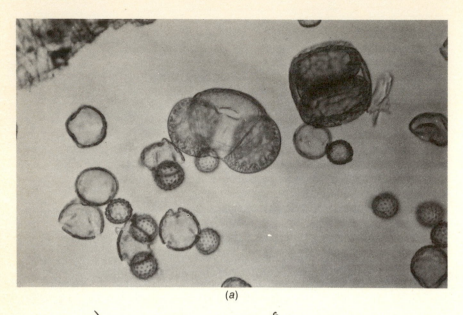

(a)

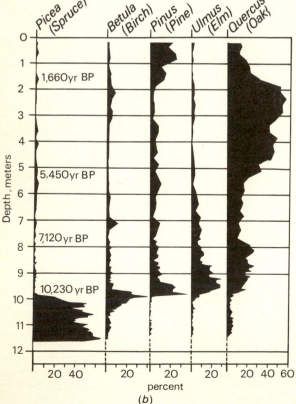

(b)

**Figure 5-3** (a) Pine, oak, ragweed, and sour dock pollen, typical of material accumulating today in small lakes in southern Wisconsin. The spherical, spiny ragweed grains are about 19 $\mu$m in diameter. *(Photo by L. J. Maher, Jr.)* (b) This partial pollen diagram shows the abundance of pollen for several tree species in bog sediments near St. Paul, Minnesota. The decrease in spruce and increase in oak pollen through time is interpreted as a change from cool and moist to warmer and drier climate. *(Modified from a pollen diagram for Kirchner Marsh, Minnesota, by T. C. Winter.)*

Another wind-deposited sediment is *loess*, composed mainly of silt-size particles. Large areas of North America are blanketed by this material as the result of deposition in the past. The sources of the loess were probably the great braided streams fed by the melting glaciers and the newly exposed glacial sediments on the uplands. Thick deposits occur along the courses of the Mississippi and Missouri rivers. The fact that these deposits become thinner toward the east indicates that the prevailing wind direction was from the west. Loess deposits are associated with every period of glaciation.

Associated with loess and windblown sand deposits in many places are polished and faceted stones called *ventifacts*. A layer of ventifacts typically forms on the surface of wind-eroded rocky material concentrated by the removal of finer-grained sediments. Sandblasting shapes them, and eventually they become buried. The association of such stones with sand and silt deposits proves their origin as wind-produced features.

## PLUVIAL LAKES

One of the most remarkable records of climatic changes is found in the desert basins of the Southwestern United States. Sediments, fossils, and

**Figure 5-4** This fossil ice wedge, developed in glacial drift near Iowa City, is a relic of a much colder past climate. The light area, now composed of fine sand and silt, was originally occupied by perennial ground ice. Subsequent melting allowed fine sediment to be washed in to fill the void.

**Figure 5-5** The elongate hills on this topographic shaded relief map are ancient sand dunes that have been stabilized by modern vegetation resulting from a shift to more moist climatic conditions. *(From Ashby, Nebraska, 15-minute quadrangle, United States Geological Survey.)*

landforms indicate that in the past these basins were permanent freshwater lakes of great extent (Fig. 5-6a). Such lakes formed when cooler temperatures prevailed, lowering evaporation rates. At the same time, higher precipitation rates furnished enough runoff to maintain high water levels. These periods of lake development are called *pluvial* intervals, and they coincided with the expansion of glaciers elsewhere.

Ancient shoreline features on the sides of the enclosing slopes mark the position of former water levels. Wave-cut cliffs, beaches, deltas, and spits are situated well above the dried floors or salty shallow remnant lakes of today (Fig. 5-6b). Extensive study of the lake sediments and water-level indicators

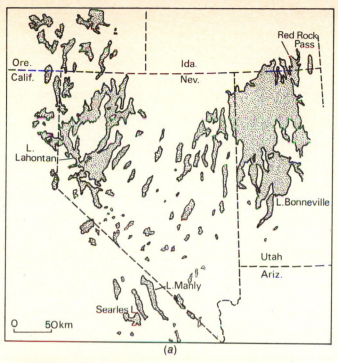

(a)

(b)

**Figure 5-6** (a) Distribution of pluvial lakes in the Southwestern United States. Great Salt Lake is a remnant of Lake Bonneville. *(Redrawn after fig. 1 from R. B. Morrison, "Quaternary Geology of the Great Basin," in H. E. Wright, Jr. and D. G. Frey (eds.), Quaternary of the United States. Copyright © 1965 by Princeton University Press, p. 266. Reprinted by permission of Princeton University Press.)* (b) A high shoreline of Lake Bonneville developed on the side of Steep Mountain, south of Salt Lake City. Flat benches such as this may be produced by wave erosion, or they may be depositional features.

reveals a history of periodic lake formation and lake drying linked to the expansion and retreat of glaciers in some of the surrounding mountain ranges, especially the Sierra Nevada of California and the Wasatch and Uinta ranges of Utah.

The largest of the pluvial lakes is called Lake Bonneville. Located mainly in the Great Basin of Utah, the lake attained a maximum size of over 50,000 km² and a depth of at least 335 m, as indicated by the highest shoreline features. Great Salt Lake is a remnant of this once great body of water. At its highest levels, Lake Bonneville spilled water through an outlet at Red Rock Pass in southern Idaho. The first spillover was in the form of a catastrophic flood that discharged into the Snake River. This Bonneville Flood is documented by erosional scars and deposits of enormous quantities of boulders along its path.

It is difficult to imagine, especially for the summer visitor, that Death Valley was ever anything more than a barren salt flat. But discernible through the shimmer of heat waves are the wave-cut benches marking the shorelines of Lake Manly, which attained a depth of over 180 m.

## CHEMICAL EVIDENCE FOR TEMPERATURE CHANGE

Many invertebrates living in the sea grow shells composed of calcium carbonate ($CaCO_3$). The elements necessary to construct the shells are derived from dissolved material in seawater. The abundance of two isotopes of oxygen, $O^{18}$ and $O^{16}$, in marine shells is related to the amounts dissolved in seawater when the shells were secreted. Changes in the ratio $O^{18}/O^{16}$ reflect changes in the volume of glacier ice on the planet in the following way. Molecules of $H_2O^{16}$ are selectively evaporated from the ocean surface, leaving the water enriched in $H_2O^{18}$. If the $H_2O^{16}$ is locked up in glacier ice during a period of cold climate, then the ratio $O^{18}/O^{16}$ increases. Conversely, when glacier ice melts, the $H_2O^{16}$ is returned to the oceans and the isotopic ratio becomes smaller.

A valuable record of such fluctuations is contained in the sediments of the deep ocean basins where remains of dead organisms become incorporated into the bottom muds. Oceanographers have retrieved a great many cores of these sediments, some spanning several million years of time. Isotope analysis allows the construction of a glacier ice volume curve, and this in turn can be used to deduce the pattern of past climatic changes.

Another long record of ancient climates is locked into the thick ice sheets of Greenland and Antarctica. It has been found that the $O^{18}$ content of precipitation at high latitudes is controlled by the temperature at which condensation took place. Studies show that high $O^{18}$ concentrations in rain and snow indicate warmer temperature of formation, whereas lower amounts mean colder temperatures. The Greenland ice sheet represents the annual accumulation of snow through thousands of years. In an ice core obtained from northwest Greenland a record of temperature fluctuation was constructed by analyzing the oxygen isotope composition of closely spaced samples of the ice. Similar studies on an ice core from Antarctica confirm the pattern of temperature changes represented by the $O^{18}$ content of the Greenland ice, indicating a contemporaneous worldwide fluctuation.

## MARINE ORGANISMS

Several species of planktonic foraminifera, microscopic organisms that live in the surface waters of the ocean, are sensitive to temperature. The relative abundances of "cold" to "warm" species in deep-sea sediments should reflect climatic conditions existing when the plankton lived. A very useful relationship between coiling patterns in the shells of certain species and temperature has also been discovered. Coiling direction is determined by noting the direction (clockwise or counterclockwise) in which chambers have been added when viewed from a specified position. The foraminifer *Globorotalia truncatulinoides* is extremely suited to such studies because the left-coiling form is tolerant of higher temperatures than the right coiling form. Climate curves are then constructed on the basis of the coiling ratios (Fig. 5-7).

**Figure 5-7**  (a) *Globorotalia truncatulinoides*, planktonic foraminifera from deep-sea sediment. Left-coiling forms on left side of photo, right-coiling on the right. Magnification 60X. *(Photo courtesy of Goesta Wollin.)* (b) A climatic curve based on the abundance of left-coiling and right-coiling forms of *G. truncatulinoides*. The data, from Core V12-122, taken in the Caribbean, show evidence of three intervals of mild climate during the Wisconsin Glaciation. *(Modified from G. Wollin and others, 1971, in Late Cenozoic Glacial Ages, p. 208.)*

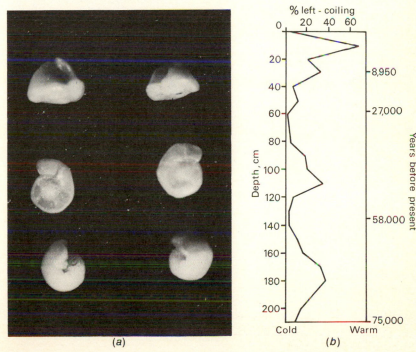

(a)                                (b)

## SUMMARY OF CLIMATIC TRENDS

Because the continental record is so diverse and poorly dated, as well as susceptible to large gaps resulting from erosion and nondeposition, the most reliable continuous imprint of past climates is contained in the marine record, including sea-level changes. The contrast is well stated by Maurice Ewing[1]: "The problem of estimating the number of glaciations from evidence on continents may be compared in complexity to estimating the number of times a blackboard has been erased; estimating the number of glaciations from evidence in deep sea sediments may be compared to finding the number of times the wall has been painted." However, oxygen isotope studies of the thick ice sheets hold promise for an extended land record in northern latitudes. Difficulties in determining real ages for sedimentary sequences have led to differences of opinion in dating major fluctuations. Even so, the great variety of different investigations that show similar patterns of change prove beyond doubt that the climate has fluctuated between warm and cold cycles. Superimposed on long-term cycles are shorter periods of change. A major research effort is presently involved in wide-ranging studies in order to recognize the periodicity and direction of climatic changes that will allow prediction of future climatic regimes. Are we experiencing the end of an Ice Age, with warmer climates to come, or is the earth entering the first cooling stage of a new glacial period? The answer to this question is not far off in the future of paleoclimatic research.

[1]Maurice Ewing, 1971, The late Cenozoic history of the Atlantic Basin and its bearing on the cause of ice ages, *in* K. K. Turekian, ed., *The Late Cenozoic Glacial Ages,* Yale University Press, New Haven, Conn., p. 572.

# SIX

## TWO MILLION YEARS OF COLD SPELLS AND WARM SPELLS

The Ice Age in North America was dominated by the periodic growth and decay of two great ice masses. The largest of these, the Laurentide ice sheet, was centered over Hudson Bay. A smaller complex of glaciers, the Cordilleran, covered the high mountain ranges of the northwest coast, stretching from Montana and Washington northward into the Aleutian Islands (Fig. 6-1). Greenland maintained an ice cover slightly larger than the present one. At the same time that the larger ice sheets were expanding, many smaller glaciers formed in the western mountains. The Appalachians south of the Laurentide ice border were apparently not high enough to spawn glaciers, even with the much lower snow line that prevailed. Much of the landscape of the continent owes its present character to the periodic advance and retreat of these glaciers, either directly from having been ice-covered, or indirectly from the effects of the glaciers on climate, surface drainage, and other geological processes.

### THE LAURENTIDE ICE SHEET

The growth of the Laurentide ice sheet was the most spectacular glaciological event in all the millions of years that make up the Cenozoic Era. Imagine the enormity of an ice mass that covered over 13 million km² of the North American continent. It contained about 25 million km³ of ice, representing enough water to lower the level of the sea by 75 m. The accumulation of such

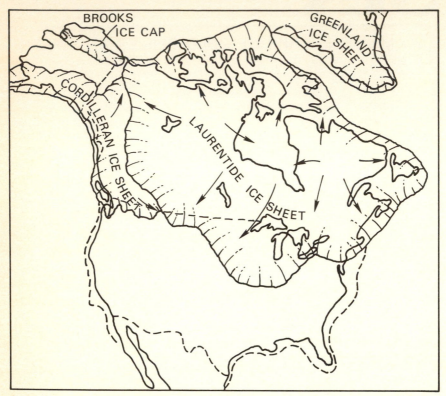

**Figure 6-1** Approximate maximum extent of major ice sheets in North America during the Great Ice Age. Ice caps and glaciers in the mountains of the Western United States are not shown. Dashed line is approximate coastline during full-glacial development.

a great volume of ice undoubtedly required many thousands of years. However, detailed knowledge of the early history of its expansion is not well known because, in the very process of spreading, the ice sheet literally covered its own tracks by subsequent erosion and deposition.

Because of the nature of the glacier-forming process, the first response to the general cooling at the start of a glacial period was snow accumulation on elevated lands and in areas of high precipitation. Northeastern Canada, especially the Labrador-Ungava Plateau and Baffin Island, afforded the necessary physical environment. Once an ice sheet has developed, it can effectively induce its own expansion by acting as a barrier to air masses. Moist air, forced to rise over the chill hill, releases precipitation as snow on the windward side, thus causing expansion of the ice mass. Boundaries of the Laurentide ice sheet at its maximum extent were fixed by calving rates along eastern North America, where its terminus lay in the Atlantic Ocean, and by melting rates along its southern and western boundaries.

**Glaciological Activity**   Very little is known in detail about how ice sheets behave. Glaciologists have made very few observations at the base of such large glaciers, and just a few deep probes have been attempted because these geological features exist today in hostile environments, such as Greenland and Antarctica. The features left behind after the disappearance of the Laurentide ice sheet indicate that the interior part is generally an area of erosion, whereas the outer margins of the ice sheet are marked by deposition. The outer boundary of erosion for the Laurentide ice sheet during its last expansion is marked by an arcuate belt of large lake basins and lowlands (Fig. 6-2). Beginning with Great Bear Lake in the Northwest Territories, this lake border includes Great Slave Lake, Lake Athabaska, Lake Winnipeg, and Lake of the Woods. The eastern segment includes the Great Lakes and the St. Lawrence River Lowland. Inward from this border, there is much exposed bedrock and mineral exploration is relatively easy. Beyond that border, the glacial sediments are much thicker, and the landscape is comprised of moraines and till plains. In this zone of deposition, the bedrock is exposed in relatively few places, making geological mapping and mineral exploration difficult.

**Figure 6-2**   Generalized regional zones of glacial activity. A major control on erosion at the base of a glacier is its temperature. The belt of maximum erosion should coincide with a basal zone very near the melting point of ice, where melting and refreezing allow the incorporation of rock debris into the glacier.

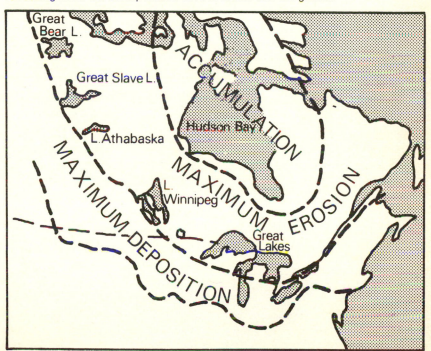

**Southern Boundaries**   Four major expansions of the Laurentide ice sheet are indicated by the existence of the deposits left behind in superposition by the successive advances. Interglacial periods are indicated by buried soils and other nonglacial geological indicators. The drift borders of these deposits are remarkably coincident, suggesting that the glacial stages were all of comparable severity (Fig. 6-3). Details of the two older glacial periods—the Nebraskan and Kansan—are lacking because weathering, erosion, and deposition, including the effects of later glacial activity, have destroyed or buried much of the geological record. The Wisconsin Glaciation was so recent, however, that a fairly detailed account of the ice sheet has been constructed. The ice sheet reached its maximum southern extension during the Illinoian Glaciation when a large protrusion crossed the 36th parallel.

Digitations in the border of the Wisconsin drift seem to be related to the underlying bedrock surface. Fingers of ice followed low trends, resulting in a preferential southward extension of the glacier terminus. These marginal ice streams are called *lobes*, and they have been given formal names where their activity is well documented. The James Lobe in South Dakota, the Des Moines Lobe in Minnesota and Iowa, and the Michigan Lobe in Wisconsin and Illinois are representative examples of such ice streams along the southern border of the Laurentide ice sheet during the Wisconsin Glaciation. Thus, the ice margin was not a great straight wall of ice but rather an irregular border of fluctuating lobes.

A peculiar interruption in the generally continuous cover of glacial sediments in the Great Lakes states is called the Driftless Area. Situated

**Figure 6-3**   General locations of the boundaries of four major drift sheets in central North America. *(From "Glacial Map of the United States East of the Rocky Mountains," Geological Society of America.)*

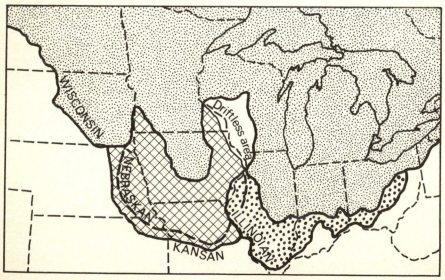

mainly in southwestern Wisconsin and surrounded by glacial deposits, the Driftless Area is a beautiful landscape of stream-dissected sedimentary rocks of Paleozoic age. None of the characteristic features of glaciation is present, such as moraines, lakes, erratics, and striated bedrock surfaces. One conclusion is that the area has never been glaciated. Whatever prevented ice from invading the area is not known. Another view, more widely held, is that glacier ice did cover the area, but that subsequent erosion by running water has removed all the evidence. The deep valley of the Mississippi River bordering the Driftless Area and the nature of the easily eroded rock types lend support to the latter view.

**Glaciation of Eastern North America**   The end moraines of the Laurentide ice sheet marking its farthest eastward advance have been drowned by the postglacial rise in sea level. Ice flow was generally easterly and southerly across eastern Canada, and the terminus was eventually controlled by rates of calving at the edge of the continental shelf. An extensive submerged moraine off the coast of Nova Scotia may mark the eastern boundary during the Wisconsin Glaciation. Perhaps the entire continental shelf north of Long Island was glaciated at one time or another.

All New England was so vigorously glaciated by the last advance that little evidence of former glaciations remains. The regional southeastward movement of the ice is demonstrated by both erosional features, especially whalebacks and striations, and by trains of boulders scattered along the glacier's course downstream from distinctive bedrock sources. The fact that even the highest mountains were completely ice-covered gives some measure of the thickness of the Laurentide ice sheet. Mount Washington in New Hampshire has an altitude of 1,905 m, and Mount Katahdin in Maine, 1,596 m. The summits of both mountains were beneath the surface of the ice sheet during the last glaciation. Therefore, the ice was nearly 2 km thick.

In several places evidence for multiple glaciation is displayed in stratigraphic sections of glacial drift. The sea cliffs along the western part of Martha's Vineyard contain tills thought to represent every one of the four major glaciations defined from studies in the Midwest. Other evidence of older glacial activity has been found in the Boston area, on Cape Cod, and in other New England localities, as well as on Long Island.

On the basis of radiocarbon ages, the last ice sheet in New England is thought to have advanced about 20,000 years ago. That advance is represented by moraines on Long Island, Cape Cod, and Martha's Vineyard (Fig. 6-4). Recessional positions and readvance constructed moraines along the coast of Maine between 14,000 and 12,000 years ago.

**Eastern Great Lakes Region**   The margin of the Laurentide ice sheet is better known west of Long Island because it was based on land, in contrast to its terminus to the east in the Atlantic Ocean. Along this margin the flow of ice was strongly influenced by topographic low trends in the bedrock, which directed its movement. East of the Adirondack and Catskill Mountains, ice flowed essentially southward along the Hudson-Champlain Valley. A much larger protrusion of ice moved out of the Ontario Basin to cover the

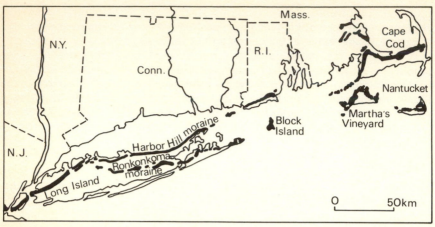

**Figure 6-4** End moraines in southeastern New England. Many are traceable below sea level on the continental shelf. *(Modified from "Glacial Map of the United States East of the Rocky Mountains," Geological Society of America.)*

Appalachian Plateau. The ice was thick enough to cover the highest peaks in the Adirondacks (greater than 1,600 m) and Catskills (1,275 m). Because drift only of Wisconsin age is found throughout most of the region, little is known of the glacial history before that time. The farthest advance reached into northeastern Pennsylvania and northern New Jersey.

One of the most impressive activities of the Laurentide ice sheet in this part of North America was the erosion of a remarkable string of lakes south of Lake Ontario. These Finger Lakes consist of 11 elongate troughs that trend generally in a southerly direction. The largest, Lake Cayuga, is over 70 km long with a greatest depth of 130 m. The bottom elevation of Lake Seneca is almost 60 m below sea level. The lakes themselves formed in the eroded troughs about 12,000 years ago when the ice sheet had melted back into the Ontario Basin.

Farther west the Laurentide ice sheet made advances deep into the center of the continent by the southward extension of several lobes of ice. A drift sheet of Illinoian age lies generally outside the moraines of the younger, Wisconsin Glaciation. Near Toronto, on the northwest shore of Lake Erie, a thick sequence of lake sediments between tills is considered to represent a warm period between the Illinoian and Wisconsin Glaciations. Pollen grains of sweet gum, a tree now growing on the Coastal Plain of the Southeastern United States, suggest a warmer climate for Toronto during the interglacial period than that of today.

Beginning about 70,000 years ago, the Laurentide ice sheet expanded southward across the St. Lawrence Lowland into the eastern Great Lakes basins (Fig. 6-5). A remarkably good stratigraphic record of tills separated by nonglacial sediments, such as peat and lake silts, indicates that the ice margin advanced and retreated several times in the following millenia. Two advances reached central Ohio and Indiana, one about 55,000 years ago and another 20,000 years ago. The intervening time is one of glacier shrinkage. One last readvance into the Great Lakes basins at about 13,000 years marked the last pulse of the dying ice sheet.

**Western Great Lakes Region and the Midcontinent**  Here the record of glacial activity is so extensive that it allowed the original recognition of the multiple nature of Ice Age climatic change. The oldest glacial sediments are found in Nebraska, representing two ice advances. Following this Nebraskan Glaciation, a long warm interval resulted in the development of deep soil on the tills. During the Kansan Glaciation that followed, ice from the Laurentide field reached its farthest westward extension into Nebraska and Kansas. The subsequent weathering of these deposits was so intense during the following interglacial that most of the rock and mineral fragments in the upper few meters were completely decomposed. The resulting soil zone, composed of sticky clay, is called *gumbotil*.

As the glacial events become more recent, their activity is better-documented because the sedimentary record is more complete. During the Illinoian Glaciation, ice reached southern Illinois, the southernmost extent of any Pleistocene continental glaciers in the Northern Hemisphere. A major path was southward along the Lake Michigan Basin. Three major advances are known from deposits in Illinois. A long, warm interglacial resulted in the

**Figure 6-5**  Major ice lobes of the eastern Great Lakes region active during the Wisconsin Glaciation. Arrows indicate general direction of ice movement. *(Modified from A. Dreimanis and R. P. Goldthwait, 1973, in The Wisconsinan Stage, Geological Society of America, Memoir 136.)*

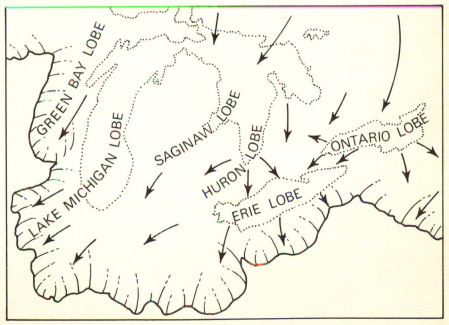

widespread formation of soils on top of the Illinoian drift. This *Sangamon soil* has been recognized throughout a wide area in the midcontinent.

All the events so far mentioned are beyond the limits of radiocarbon dating. Therefore, the exact time of their occurrence is not known. A layer of volcanic ash found in association with drift of Kansan age gives a date of 600,000 years ago, based on fission-track dating. Present estimates indicate that the Sangamon soil began to form about 100,000 years ago.

Glaciers of the Wisconsin Glaciation are well known from the impressive system of moraines that garland the midcontinent landscape (Fig. 6-6). The

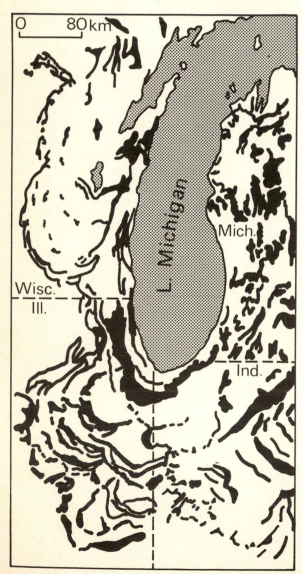

**Figure 6-6** End moraines of the Wisconsin Glaciation. Many of these landforms mark the successive positions of retreating glacier margins; the outermost moraines represent the farthest ice advance. *(From "Glacial Map of the United States East of the Rocky Mountains," Geological Society of America.)*

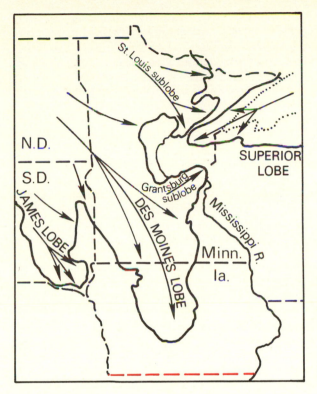

**Figure 6-7** Ice lobes active in the Midwest during the last part of the Wisconsin Glaciation. The Des Moines Lobe reached its maximum extent about 14,000 radiocarbon years ago.

southern ice margin fluctuated many times during this last 70,000 years of the Pleistocene Epoch, resulting in the constant shifting of climatic zones. The Superior and Michigan Basins acted as reservoirs of ice from which periodic glacier "floods" emanated.

Stratigraphic studies and radiocarbon dating indicate that glaciers were active in Illinois, Wisconsin, Iowa, Minnesota, and the Dakotas many different times during the last 70,000 years. A major warming took place between 30,000 and 20,000 years ago, and the most recent warm spell set in about 14,000 years before the present. An ice lobe advancing from Green Bay left a remarkable swath of drumlins along its path to Madison, Wisconsin. In Minnesota two large ice lobes were active. One periodically moved southward from the Superior Basin, and the other followed a low topographic trend from the Dakotas into Iowa (Fig. 6-7). The latter reached its maximum extent at Des Moines about 14,000 years ago. In the western Dakotas, the drift border follows a northwestern course, indicating a sharp climatic control on the ice-sheet terminus.

## CORDILLERAN GLACIER COMPLEX

The high western mountain ranges of North America became the spawning grounds for hundreds of local glaciers in response to the lower temperatures of the Ice Age. At first, valley glaciers expanded in all directions from the Cascades and Coast Ranges and from the high ranges of the Northern Rockies. Eventually these glaciers coalesced to form the continuous Cordilleran ice sheet from Puget Sound northward into the Aleutian Island arc and stretching eastward to meet the Laurentide ice sheet (Fig. 6-1). A much smaller ice cap formed over the Brooks Range in Alaska, leaving a broad ice-free corridor in the lowland drained by the Yukon River. Other significant ice fields and expanded glacier systems were located in the Klamath and Cascade ranges of Oregon, the Sierra Nevada in California, and the Central and Southern Rocky Mountains. The total ice cover for the entire Cordilleran glacier complex approached 2 million km², only about one-fifth as large as the Laurentide sheet, but still an impressive ice cover.

A remarkable consequence of the activity of the Cordilleran ice sheet was the periodic damming of major river valleys by ice lobes pushing southward from Canada into Montana, Idaho, and Washington (Fig. 6-8). Behind these ice dams enormous lakes became ponded. Each damming episode was followed by the collapse of the ice dams as the Cordilleran ice sheet retreated, resulting in the catastrophic discharge of floodwaters across the landscape. The largest of these glacial lakes, Lake Missoula, rose periodically behind an ice lobe that plugged the Clark Fork River Valley near the Montana-Idaho border. At peak stage Lake Missoula had a surface area of about 7,500 km² and contained an estimated 2,000 km³ of water. All this water is thought to have discharged westward in a matter of a few days when the dam failed about 22,000 years ago. This great flood moved boulders with diameters greater than 10 m and scoured a system of *coulees* across the Columbia Plateau. This great tract of flood-eroded topography is called the *channeled scablands* (Fig. 6-8).

Repeated scouring by the Cordilleran glaciers changed the character of the mountain landscapes dramatically. Rock basins called *cirques* were eroded deeply into the mountain sides wherever the snow line allowed glacier formation (Fig. 6-9a). Growth and spillover from the cirques sent rivers of ice to fill valleys cut previously by streams, reshaping them into deep, flat-bottomed troughs such as the Yosemite Valley in the Sierra Nevada of California (Fig. 6-9b). Glacier deepening left smaller tributaries hanging high on the valley sides, instigating waterfalls. Sharp peaks and ridges were chiseled from the original mountain mass by the abrasive effects of glaciation. Deposition left garlands of stony sediments in the form of moraines along the ice margins, and many beautiful lakes have since formed behind the moraine dams (Fig. 6-10).

Studies in many mountain regions of the West confirm a complex sequence of multiple glaciation. Radioactive dating of volcanic ash and lava flows interbedded with glacial deposits in the Sierra Nevada indicates that glaciers were active there beginning as early as 3 million years ago. Probably the major episodes were coincident with the expansion of the larger ice sheets. The main Cordilleran ice sheet centered in British Columbia spread drift during at least four major expansions.

The last major shrinkage of the Cordilleran glaciers began about 15,000 years ago; however, the high mountain ranges afford a sanctuary today for

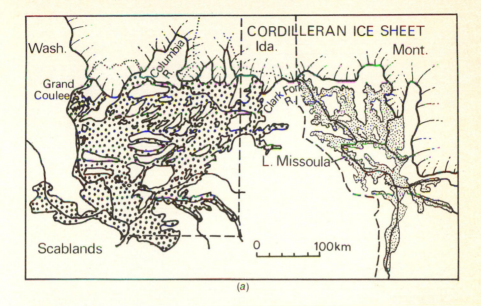

(a)

(b)

**Figure 6-8** (a) Periodically during the Ice Age, glacier tongues from the Cordilleran ice sheet dammed valleys to form large lakes. One of the largest, Lake Missoula, provided enormous floods that eroded the channeled scablands in eastern Washington. *(Modified from V. R. Baker, 1973, Geological Society of America Special Paper No. 144.)* (b) Air view of giant potholes and channels carved by the turbulent floodwaters produced by the catastrophic draining of Lake Missoula.

(a)

(b)

**Figure 6-9** (a) This cirque is a rock basin eroded at the head of a glacier, Canadian Cascades near Vancouver, B.C. *(Photo by D. A. Rahm.)* (b) Yosemite Valley, a classical glacier-eroded trough. Notice the small hanging valley with waterfall. *(Photo by R. W. Ojakangas.)*

**Figure 6-10**   Newfork Lake, Wind River Mountains near Pinedale, Wyoming, is enclosed by moraines of Wisconsin Age. Note the recessional moraine that segments the lake, and the lateral moraines of different height on the left. *(Photo by L. J. Maher, Jr.)*

many smaller glaciers and ice fields. These remnants of the great Cordilleran Complex still act as sensitive indicators of climatic change. Their fluctuations during the past few thousand years document continuing cycles of change. Perhaps, with a recognition of the rhythms of past climatic change, long-range predictions for the future will furnish invaluable warnings of impending climatic deterioration.

## BEYOND THE ICE

Outside the ice border, profound changes in the landscape were effected by the colder climates of the glacial periods as well as by the physical presence of the ice sheets themselves. Some of these effects have already been mentioned, such as sea-level changes, the lowering of ocean temperatures, and the formation of pluvial lakes in the western basins. A great many other changes in geological processes and landscape aspect that accompanied the growth of glaciers are documented in the geologic record.

**Streams**   The immediate response of rivers to the drop in sea levels accompanying glacier growth was to deepen their valleys in the course of adjusting to the new lower base level of erosion. As a result of downcutting, the former floodplains of many rivers were left high and dry as *terraces*. H. N. Fisk recognized four cycles of valley cutting and subsequent backfilling in the lower reaches of the Mississippi River that he ascribed to a fall and rise of sea level accompanying each of the major glacial periods. Farther upstream, rivers that directly drained the melting ice margins were also presented with large quantities of glacier-derived sediments for transport. Thus, at the same time that the lower Mississippi was cutting a deeper valley to meet the lower sea level, the upper part of the river valley was filling up with excess sediments it did not have the competence to transport (Fig. 6-11).

**Figure 6-11** The reaction of the Mississippi River to various changes during a glacial-interglacial interval.

| Mississippi river upstream near glacier margin. | Mississippi river downstream near base level |
|---|---|

### Glacial Period

| | |
|---|---|
| 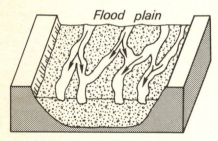 | 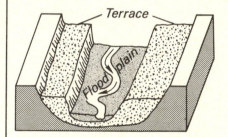 |
| (a) Meltwater streams, choked with glacial sediments, fill old valley and build a broad floodplain. | (a) River erodes through old valley fill to meet lower sea level. |

### Interglacial Period

| | |
|---|---|
| 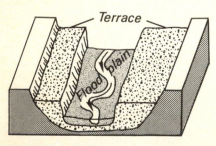 | 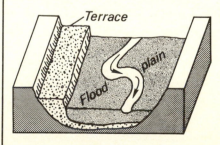 |
| (b) Valley is cut as glacial load decreases and drainage basin area expands during glacial retreat. | (b) Valley backfills in response to rise in sea level. |

Rivers not so directly linked to glacial activity responded to increases or decreases in discharge and load by cutting or filling their valleys. However, it is sometimes difficult to reconstruct the exact hydrologic and geologic factors that triggered the changes. In many parts of the continent cooler, wetter climates resulted in larger discharges that widened and deepened many river valleys. Small rivers that today occupy such large valleys are said to be *underfit streams.*

**Loess**  Rivers carried glacial outwash far beyond the ice margins, in the form of braided valley trains. These alluvial surfaces acted as sources for fine-grained sediments to be eroded by wind. The midcontinent of North America is blanketed by a deposit of loess derived primarily from the Mississippi and Missouri river systems (Fig. 6-12), when they discharged

**Figure 6-12** (*a*) Generalized distribution of loess, and the relationship of this wind-blown deposit to the major river valleys that carried meltwater and outwash from the glacial margins. *(Modified from M. M. Leighton and H. B. Willman, 1950, Journal of Geology, vol. 58.)* (*b*) Roadcuts in loess in Vicksburg, Mississippi. The silt-sized loess particles are cohesive enough to stand in vertical slopes, making excavation work safe and easy.

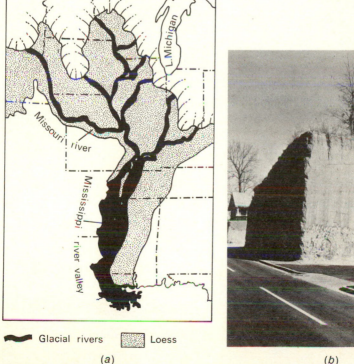

Glacial rivers  Loess

(*a*)  (*b*)

glacial meltwaters and sediment. Thicknesses of more than 30 m accumulated to the lee of prevailing westerly winds. The great dust storms that must have transported this large quantity of sediments no doubt rivaled in magnitude the storms of the 1930s that centered about the "dust bowl" of the Great Plains. The soils of much of the interior lowlands owe their richness to this fine-grained mineral "parent" material.

**Frost Action.**　　Much of the terrane near the glacier margins was subjected to temperatures low enough to result in the development of permanently frozen ground, and frost action was accelerated. Many localities south of the drift border display relics of this cold condition in the form of sorted patterned ground features not being actively constructed under present climatic conditions. Frost shattering produced great rubble fields in the Appalachian Mountains and other areas of exposed bedrock. Downslope movements were enhanced by the presense of permafrost and ground ice. Typical of the rubble deposits resulting from the colder climate of glacial periods is the Blue Rocks block field in southeastern Pennsylvania (Fig. 6-13), 65 km south of the Wisconsin drift border.

**Figure 6-13.**　　Blue Rocks block field in Berks County, Pennsylvania, was produced by frost shattering and downslope movement under climatic conditions much colder than the present. (*Photo by Noel Potter, Jr.*)

In the midcontinent, fossil ice-wedge casts (Fig. 5-4) in Wisconsin, Ohio, Iowa, and Illinois indicate a mean annual temperature as much as 10°C below that of today. One remarkable occurrence near DeKalb in north-central Illinois is a swarm of more than 500 circular mounds up to 1 km in diameter. The hills are believed to be the remnants of dome-shaped mounds pushed up by the pressure of large lenses of ground ice that formed in frozen soil.

**Vegetation**   In response to the advancing ice sheets and deteriorating climates of glacial periods, plant and animal communities migrated away from areas of stress into regions conducive to their survival and regeneration. These migrations were generally southward and toward lower elevations in highland regions. Certain areas afforded refuges closer to the glaciers, such as the Driftless Area in southwestern Wisconsin during the last glaciation and the central interior of Alaska.

A general pattern of vegetational zones that prevailed at the time of the maximum advance of Wisconsin ice sheets might serve as a model for the earlier glaciations (Fig. 6-14). The study of fossil pollen grains, combined with radiocarbon dating, indicates that the borders of modern vegetation zones were shifted as much as 1,000 km south of their present limits. Evidence comes mainly from analyzing the pollen sequences in lake and bog sediments. The Laurentide ice sheet was bordered by a narrow belt of treeless tundra, probably stretching from southern New England westward all the way to the Rockies. South of the tundra a broad belt of coniferous forest consisting of spruce and pine covered much of the Great Plains. Eastward it extended from Illinois across the Appalachians to the Coastal Plain. The southern boundary of this great boreal forest is not well established; however, it may have been as broad as 1,000 km. In mountain regions

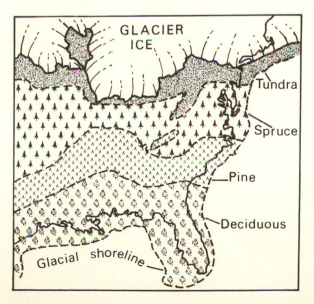

**Figure 6-14** Possible pattern of vegetation in the Eastern United States during the maximum advance of the Laurentide ice sheet during the Wisconsin Glaciation. *(Adapted from D. R. Whitehead, 1973, Quaternary Research, vol. 3.)*

the tree line was as much as 1,000 m lower, and many of the higher peaks were ice-covered, especially in the West. Vegetation expanded across the foothills and lowlands of the desert basins, as indicated by the presence of pine, spruce, and fir trees where now only desert shrubs grow.

# SEVEN

AFTERMATH OF THE GLACIAL ORDEAL

About 18,000 years ago the climatic conditions that favored the growth of continental glaciers began to change, and the expansion of glaciers across North America stalled. The cold white world of the Wisconsin Glaciation had reached a climax, and the general destruction of the Ice Age glaciers was imminent. The succeeding thousands of years right down to the present are marked by a general shrinkage of glaciers throughout the world. Accompanying the warming trend was a complicated sequence of geologic and biologic events, all intertwined in cause and effect as the earth's surface adjusted to the melting of the glaciers. Change is the name of the geological game, and this latest chapter in the history of the earth exemplifies that fact in an extraordinary way.

Enormous quantities of water, stored as glacier ice for many thousands of years, returned to the ocean reservoirs via meltwater streams draining the retreating ice margins. As a result, sea level began to rise, once again flooding the continental margins. The crust of the earth itself, which had been depressed under the weight of thousands of meters of ice, began to spring back to former heights as the glacier load was removed. Changing weather patterns triggered a variety of adjustments in geological processes and vegetational distribution. Most dramatic of all, a new landscape slowly emerged from beneath the cover of ice to become part of the familiar face of North America we know today.

## GLACIER RETREAT

The general warming trend in world climate resulted in an increase of glacier melting over accumulation, and the great ice sheets began to lose mass. This negative economy was reflected in both a thinning of the ice and a general recession of the margins. Glaciologists refer to the general withdrawal of an ice cover as a retreat; however, it must be kept in mind that no reversal of flow is involved. There are two general styles of glacial retreat. One simply involves the gradual recession of an active ice margin because the volume of ice flowing to the terminus is not sufficient to replace that lost to melting. Commonly, glacier retreat may slow or stop long enough to allow the construction of a sedimentary pile along the margin. Such *recessional moraines* are valuable aids to the glacial geologist because their spacing might allow the calculation of rates of melting if appropriate materials are available for radiocarbon dating (Fig. 7-1).

A second style of retreat involves the large-scale halt in the movement of

**Figure 7-1** Looping recessional moraines in front of the Eureka Glacier, Alaska Range, mark positions of stillstand during a general retreat of the glacier. *(Photo by Austin Post, United States Geological Survey.)*

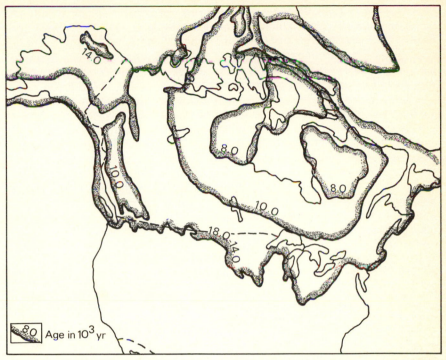

**Figure 7-2**  Position of glacier margins at various times during the general deglaciation of North America. Ages are in thousands of years. Minor readvances are not
shown. *(After V. K. Prest, 1969, Geological Survey of Canada, Map 1257A.)*

large areas of glacial ice. Such stagnant or dead ice then disintegrates and
melts in place, leaving behind a distinctive set of landforms. The slow
downmelting of stagnant dirty ice results first in the accumulation of a thin
layer of debris at the top surface of the ice. Just a few centimeters is sufficient
to act as insulation and to provide the necessary footing for vegetation to
become established. Meltwater streams cut gorges into the ice-cored topography; underground water produces systems of tunnels and caverns in the
ice, and these eventually collapse. Finally the entire mass of ice disappears,
leaving behind a complex of eskers, kames, and kettle lakes, all distinctive
features of ice stagnation processes (Fig. 3-6).

  V. K. Prest of the Geological Survey of Canada has constructed a general
model showing the positions of ice margins at various times during the
general deglaciation of North America, based on field studies of moraines
and radiocarbon age dating (Fig. 7-2). According to the available field
evidence, the Laurentide ice sheet began to deteriorate as early as 18,000
years ago along a segment from Ohio eastward, including the entire margin
along the New England and Maritime coasts. There, the ice extended far out
onto the continental shelf, and an important mechanism of loss was the
calving of icebergs. By about 13,000 years ago the margin of the ice sheet

had receded to a position parallel to the present coast of Maine. Continued wastage left most of the United States ice-free by about 12,000 years ago, except for local ice caps in Maine and lobes in the Superior and Michigan Basins. During the same period of time the Cordilleran ice sheet in the West had also retreated mainly into Canada; however, continued high snowfall maintained the major part of the glacier mass.

Even as the ice sheets were generally being depleted overall, several significant expansions of the ice margin occurred. One of these pulses sent ice lobes southward into South Dakota, Minnesota, and Iowa about 14,000 years ago. Other minor advances in the Great Lakes region are recorded at about the same time. One of the most famous Pleistocene localities in North America is the Two Creeks Forest Bed located 130 km north of Milwaukee on the western shore of Lake Michigan. Exposed in the wave-cut cliffs is evidence of a readvance of a lobe of ice in the Michigan Basin (Fig. 7-3). Till at the base of the cliff represents the main advance of ice during the last major Wisconsin glaciation. The retreat of the Laurentide ice sheet is next recorded in the overlying lake sediments, which formed in the ponded waters of Glacial Lake Michigan in front of the receding ice margin. A soil, and trees rooted in the lake sediments, represent a time of low water in the history of the lake. The readvance of ice is recorded in the stratigraphy by two distinct layers of sediment. Burying the forest and soil horizon is coarse, sandy lake sediment, representing a rising lake level as ice again moved southward, restricting the area of the basin. Radiocarbon dating indicates that the trees were killed by the high water about 12,000 years ago. Overlying the lake sediments is a till deposited directly by the advancing glacier ice. The ice that deposited the youngest till was thin and short-lived in terms of activity. Such fluctuations of the ice margins probably represent instability in the Laurentide ice sheet, produced by the general warming trend. There is good evidence that rapid movements of glaciers, called *surging*, are caused by the warming of polar glaciers to the melting point. Such an increase in temperature promotes sliding owing to the lubricating effect of water and more rapid internal movement of ice.

As both the Cordilleran and Laurentide ice sheets continued to shrink, an ice-free corridor developed in western Canada, linking the unglaciated sector of Alaska to the heartland of the continent. In the next chapter, this freeway will be discussed as the possible route of the first human migrations into the continental interior. By 8,000 years ago the Laurentide ice sheet had shrunk to two small ice caps separated by Hudson Bay. The Cordilleran ice was also segmented into several ice fields centered upon the highlands of the Coast Ranges. Elsewhere in the western mountain ranges, all the mountain glaciers were in general retreat, as the snow line climbed to successively higher altitudes. By 5,000 years ago, the glaciers of North America had probably melted down to approximately their present distribution. Since that time the ice has been mainly restricted to refuges in the high mountains, the polar regions, and areas of great snowfall. Minor climatic changes have triggered modest advances several times in the last few thousand years; however, worldwide warm temperature generally has prevented the reconstitution of the great ice sheets of the past.

(a)

Soil
Lake sediments

Till

Lake sediments
Forest bed

Lake sediments

Till

(b)

**Figure 7-3**  (*a*) Log from an ancient forest near Two Creeks, Wisconsin, buried by glacial drift about 12,000 radiocarbon years ago. (*b*) Generalized stratigraphy at Two Creeks indicates two glacial advances and several higher stands of Glacial Lake Michigan.

## LAKES ASSOCIATED WITH GLACIATION

The warm winds that brought destruction to the Ice Age glaciers also resulted in the drying up of the deep pluvial lakes in the Southwest. But farther north other large natural catchment basins were exposed in the lee of the retreating ice margins. The glaciated portion of the continent became a vast storehouse of fresh water, including some of the largest lakes ever to have existed in North America. Many of these lakes formed against ice dams created by the glaciers, and their history is intimately linked with the activity of advance and retreat of ice margins. Some of the largest of the lakes became extinct as the glacial ice dams melted, but many more have persisted right to the present.

Lakes, by their very nature, are generally fleeting geological features on any landscape for several reasons. They are natural traps, not only for water, but for sediment as well, and therefore, extinction by in-filling is a common fate. Natural plumbing systems in the form of stream networks also commonly cause the disappearance of lakes by eroding outlet streams which drain them away completely. Few lakes within the boundaries of Ice Age glaciations originated earlier than the Wisconsin Glaciation because the processes of lake destruction are so rapid and effective. In fact, a good approximation of the southern limit of the Laurentide ice sheet during its last advance in the midcontinent is a line separating areas where lakes are abundant from those where no lakes exist.

Lake basins were produced in several ways: by glacial erosion, by glacial deposition, and by a combination of those processes, along with the melt-out

**Figure 7-4** Kettle lakes are formed when glacial sediment collapses as the result of the melting of buried stagnant ice. Terminus of Nabesna Glacier, Wrangell Mountains, Alaska. *(Photo by Noel Potter, Jr.)*

of buried ice blocks. Glacial erosion was very effective in mountain areas; there, cirque basins scoured out of the mountain sides filled with water to become *tarns*. *Pater noster lakes* are strings of lake basins connected by overflow streams; they occupy valleys that were gouged out irregularly by eroding ice streams. The name comes from their pattern on topographic maps, which resembles the beads on a rosary. A third category of lakes common to glaciated mountains formed behind moraine dams constructed across the glaciated valleys (Fig. 6-10).

Continental glaciers left a deeply eroded bedrock surface in many areas as they melted, and some of the vast lake country of the Canadian Shield originated in this way. Along the retreating ice margins, irregular deposition of glacial sediments and the collapse of sediments over melting stagnant ice blocks resulted in an intensely pockmarked landscape that now contains thousands of kettle lakes (Fig. 7-4). Some very impressive lakes formed where the deposition of moraines blocked the natural drainage.

Even though meltwater from the receding ice may have been initially trapped in the newly exposed basins, the lake waters have been recycled many times during the thousands of years of postglacial time. Maximum residence time for water in even the largest basins is not more than a few hundred years.

**Lake Agassiz**   The largest of the glacial lakes to have existed in North America is named Glacial Lake Agassiz. Remember Louis Agassiz? This lake began to form about 12,000 years ago when the Laurentide ice sheet melted north of a major topographic divide in western Minnesota. As the ice continued to retreat, Lake Agassiz expanded across more than 350,000 km$^2$ in Minnesota, the Dakotas, Saskatchewan, Manitoba, and Ontario (Fig. 7-5) and attained a maximum depth of almost 200 m.

**Figure 7-5** Eventual extent of Glacial Lake Agassiz after retreat of the Laurentide ice sheet. The southern outlet eroded the impressive valley now occupied by the Minnesota River. *(After J. A. Elson, 1967, Life, Land and Water, University of Manitoba Press.)*

The history of the expansion and eventual drainage of Lake Agassiz is contained in the erosional and depositional features it left behind. Successive lake levels are documented by such features as beach sediments, wave-cut platforms, and deltas that were constructed by streams entering the lake. Deeper-water sediments consist of thick layers of silt and clay. Lake levels were controlled by the amount of water entering the lake compared to the volume lost, especially by evaporation and outlet streams. During its early history Lake Agassiz had but one outlet, Glacial River Warren, whose waters eroded the wide, deep valley now occupied by the Minnesota River. Later, lower outlets were opened as the Laurentide ice dam melted northward. By the year 8000 B.P. (before the present), the lake had lost most of its water. Remnants of Lake Agassiz are Lake of the Woods, Lake Winnipeg, and Lake Winnipegosis. During its existence, the lake was a major influence on the natural environment of the midcontinent; today the old lake bed is a vast flat plain that supports a rich agricultural economy in the Red River Basin of the North.

**Glacial Great Lakes**  To the east of Lake Agassiz another system of glacial lakes had begun to form a few thousand years earlier (Fig. 7-6). The expansion of these Glacial Great Lakes was synchronous with the general retreat of the ice margin into Canada. Even prior to glaciation, the land area now covered by the Great Lakes was low because of a long interval of stream erosion, which carved wide valleys into the bedrock. Repeated glaciation removed additional volumes of rock to form basins, some with bedrock floors well below modern sea level. The early postglacial history of these lake basins is complicated by several factors, including periodic readvances of

**Figure 7-6**  Stages in the development of the modern Great Lakes. (*a*) 13,000 years ago. (*b*) 11,500 years ago. (*c*) 9,500 years ago. (*d*) 6,000 years ago. (*After V. K. Prest, 1970, Quaternary Geology of Canada, Geological Survey of Canada.*)

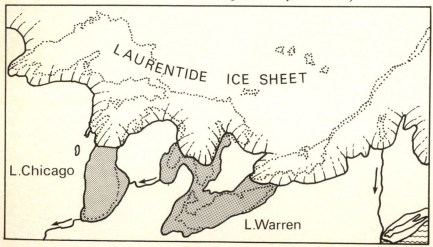

(a)

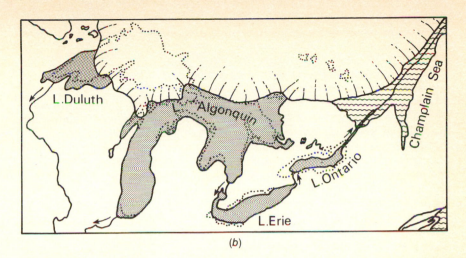

(b)

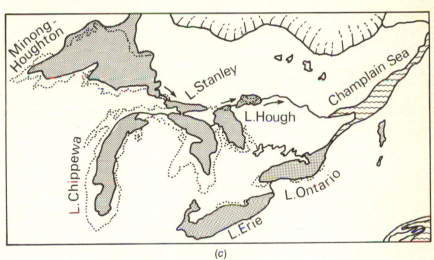

(c)

(d)

ice, vertical crustal movements resulting from the unloading of glacial ice, and erosion of lake outlets.

In general, all the Great Lakes experienced several high-water stages during their early history, resulting in an expansion of shorelines far beyond the limits of the modern lakes. These early lakes left shoreline features and fine-grained lake sediments to mark their positions. Water levels during these high stages were generally controlled by ice dams to the north in the same basins, and by the elevation of outlet streams, just as the overflow drain on a bathtub controls the maximum level to which the tub can be filled. High-water stages were followed by a second phase, characterized by lake levels lower than at present as the result of drainage through low outlets exposed during final ice retreat from the region. These outlets owed their initial low elevations to a glacially depressed crust of the earth. The surface of Lake Superior was 75 m below its present level, Lake Michigan 100 m lower, and Lake Huron 125 m below its modern stand.

A third phase in the history of the lakes was one of general expansion as water levels rose in response to vertical uplift of the outlet areas owing to rebound of the crust. Rising waters in Lakes Superior, Michigan, and Huron eventually merged into one lake, to form the largest of all postglacial Great Lakes. This Nippissing Great Lakes stage endured for about 1,000 years. Downcutting of outlets has since resulted in the lowering of the lakes to their present levels and separation into individual basins.

## POSTGLACIAL SEA-LEVEL CHANGE

The melting of the Ice Age glaciers had an immediate and worldwide effect on sea level as water was returned to the ocean reservoirs. The rate of rise is linked directly to the time involved in the disappearance of ice from the continents, and the magnitude of the increase determined by the net volume of water returned for storage. As already pointed out in Chap. 5, reconstructing past sea-level changes is a difficult endeavor because the crust of the earth has not remained stable through geologic time. Another important complication lies in the evidence that the ocean basins themselves may be growing larger as the result of sea-floor spreading, a process whereby new ocean crust is formed along oceanic ridge systems, accompanied by continental drift.

What has been the change in sea-level since the Ice Age glaciers began to melt 18,000 years ago? Several different approaches have come up with approximately the same answer. K. O Emery, of Woods Hole Oceanographic Institute, dredged up samples of ancient peat, oyster shells, and other organic shallow-water indicators from depths as great as 130 m on the continental shelf of Eastern United States. Because these fossil remains indicate a shoreline environment, their present depth of submergence is a measure of the rise in sea level since the organisms died. By determining the ages of many such discoveries and noting the water depth from which the materials were recovered, Emery was able to construct a continuous record of sea-level change over the past 35,000 years (Fig. 7-7). Because no shoreline indicators were found in water depths greater than 130 m, that figure must represent the maximum postglacial change.

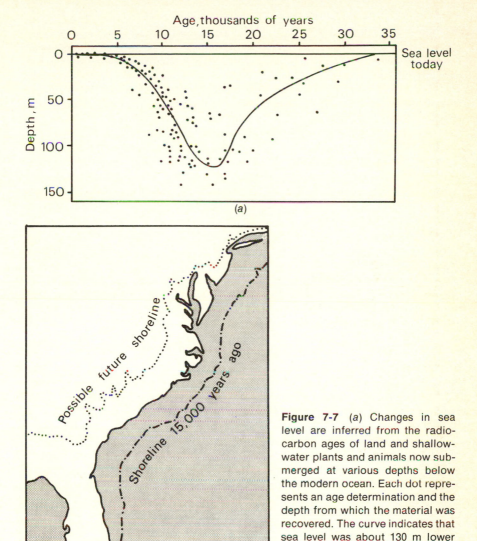

**Figure 7-7** (a) Changes in sea level are inferred from the radiocarbon ages of land and shallow-water plants and animals now submerged at various depths below the modern ocean. Each dot represents an age determination and the depth from which the material was recovered. The curve indicates that sea level was about 130 m lower 15,000 years ago. (b) Ancient and modern shoreline and a possible future shoreline if all present-day glaciers should melt. *(Data from K. O. Emery, The Continental Shelves. Copyright © 1969 by Scientific American, Inc. All rights reserved.)*

A. L. Bloom made a careful inventory of the total volume of ice lost to ablation during the last ice recession, using maps showing the position of the retreating glaciers at various times, such as the one shown in Fig. 7-1. He based his estimate on the thickness of the ice sheets (a maximum of 3.5 km for the Laurentide, for example) on knowledge of present continental glaciers in Greenland and Antarctica. Radiocarbon dating of successive retreat positions allowed the calculation of rates of melting. Conversion of ice volumes to an average ocean water depth results in a figure for the eustatic change of about 130 m.

Both studies prove that the shorelines of the glacial periods were far seaward of their present positions. If the past is also the key to the future, then the prediction can be made that if glaciers continue to melt, a future shoreline on the East Coast of the United States might occupy the position shown in Fig. 7-7. Considering the population density of coastal land, it is not difficult to predict the future necessity for massive human migrations.

Thus, the familiar features of the present coastal areas of North America became established only during the last 4,000 or 5,000 years. Wave erosion has cut cliffs and benches, and longshore currents have deposited beaches and sandbars. Coastal river valleys, cut deep to meet the lower sea level of the glacial period, have been drowned by the rising sea to form estuaries. Along glaciated coasts the deep troughs eroded by glaciers have also been flooded. These *fiords* are most characteristic of the northwestern coast of North America, where they afford a great network of inland passageways (Fig. 7-8). Chatham Strait in southeastern Alaska is the longest fiord in the world, almost 400 km, with a width of 10 km and a depth of up to 700 m. As long as sea level continues to invade the low continental fringes, changes will continue to shape the coastal regions of the world.

## POSTGLACIAL UPLIFT

One rather surprising fact about recently deglaciated terranes is that they are all areas of active vertical uplift. This condition was first suspected because along glaciated coasts shoreline features and recent marine sediments are found high above the present sea level. Clearly, either the land has experienced uplift, or sea level has been higher in the recent past. The latter possibility, however, was rejected on the grounds that sea level in fact must have been lower because of the existence of continental glaciers. In the midcontinent region it was noted that continuous shorelines of ancient glacial lakes are not level but rise generally as they are traced northward. The highest beach constructed by Lake Agassiz, for example, has an elevation of about 325 m near its southern end; 400 km north, the same beach is found at an elevation of almost 380 m, indicating a minimum vertical rise of 55 m since its formation. A great many careful studies of such raised features clearly indicate that the boundaries of uplifted areas are generally parallel to the limits of the last glacial ice cover; furthermore, maximum uplift coincides with areas inferred from other evidence to have been beneath the thickest part of the Laurentide ice sheet (Fig. 7-9). From these observations comes the conclusion that the outer shell of the earth was depressed under the weight of continental glaciers. Removal of the load was followed by crustal rebound, which continues toward recovery of gravitational balance.

Radiocarbon dating and the measurement of magnitudes of vertical

**Figure 7-8**  A fiord, coast of Alaska, south of Juneau. These deep inland protusions of the sea are drowned glacial valleys typical of many glaciated coasts. *(Photo by D. A. Rahm.)*

movement of tilted shorelines from place to place have allowed the calculation of rates of uplift in many places. Although uplift began at the time the glaciers started to thin, only that part of the movement since the areas became generally ice-free was recorded in the geologic record. Up to 275 m of vertical movement has occurred in the Hudson Bay area, resulting in the decanting of marine waters from the continental area (Fig. 7-10). Age dating shows clearly that maximum rates of uplift were experienced during the first few thousand years of recovery, with much slower rates prevailing after that time. W. R. Farrand, of the University of Michigan, has constructed uplift curves for the Great Lakes region and Arctic Canada that consistently show this relationship (Fig. 7-9*b*).

The circumstance of a depressed crust in glaciated coastal areas during the early part of the postglacial period resulted in a marine invasion of significant extent. A long arm of the sea occupied the St. Lawrence River Valley and extended into Lake Champlain (Fig. 7-6*b*). Hudson Bay was almost a third larger in areal extent. The shallow marine environment did not last

(a)

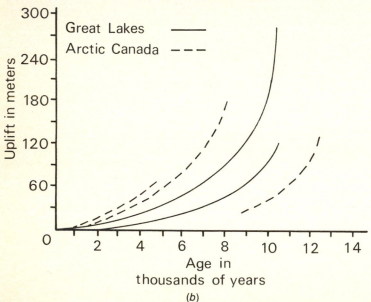

(b)

**Figure 7-9** (a) Raised beaches around Johnson Island, Hudson Bay, indicate that the earth's crust is rising in response to glacial unloading. *(Photo by J. A. Donaldson.)* (b) Rates of uplift are determined by radiocarbon dating organic materials in raised beaches, such as those in Fig. 7-9a. *(Adapted from W. R. Farrand, 1962, American Journal of Science, vol. 260.)*

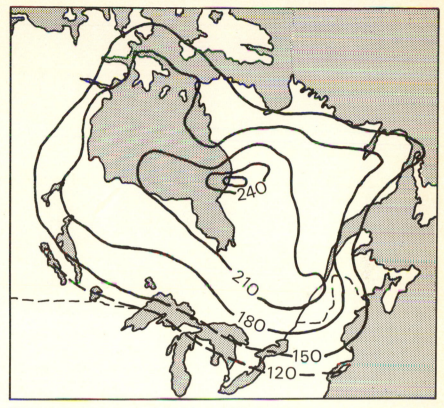

**Figure 7-10** Maximum postglacial rebound of northeastern North America, in meters. *(Data from J. T. Andrews.)*

long, however, because rapid postglacial uplift soon overtook the rate of rise of sea level, and these areas became emergent. The highest elevation on the present landscape marked by this early invasion of the sea is called the *marine limit*. Old shoreline deposits, as well as thick accumulations of silt and clay containing the remains of marine organisms, are generally found between that boundary and present sea level. During the past 5,000 years some parts of the east coast have again been invaded because slowing rates of uplift have been surpassed by the continuing eustatic rise of the sea as the glaciers of the world continue to melt.

# EIGHT

FOSSILS OF THE ICE AGE

Several years ago bulldozers cut a wide swath across swampy ground in central Minnesota in preparation for the construction of a new highway. A few meters below the surface their blades encountered a layer of muck containing the well-preserved skeletons of dozens of large, extinct bison (Fig. 8-1). The discovery of this Ice Age boneyard provided important details about the natural history of animals that had lived in Minnesota just after the last glaciers had melted. Elsewhere on the continent other evidence of former life has been found in the form of bones, tracks, excrement, plant fragments, and shells. This fossil record is the basis for reconstructing the life of former times.

The chances of preservation of plants and animals after death are generally poor because rapid burial is a requirement. Oceans provide such a service, and therefore, burial at sea has resulted in a rich fossil record for marine organisms. On the continents, where erosion is very active, natural sediment traps, such as lakes and swamps, as well as floodplain sediments and wind deposits, provide a quick, natural burial for dead plants and animals. In some places, caves not only sheltered Ice Age animals but also became mausoleums for residents that died and were buried beneath accumulating layers of cave debris. Sticky oil seeps in California snared animals, and the petroleum acted as an imperfect embalming fluid. Farther north, in the Arctic, the natural deepfreeze of frozen ground provided cold storage for the carcasses of animals, in some instances preserving skin and flesh for thousands of years.

**Figure 8-1** Skull and loose bones of *Bison occidentalis*, which is now extinct, were uncovered in a peat bog during construction of Interstate Highway 94 near Melrose, Minnesota. Radiocarbon dating yielded an age of over 7,000 years. *(Photo by K. R. Moran.)*

Even with the scant evidence provided by the imperfect fossil record, a general outline of the character and activity of Ice Age life in North America is fairly clear. Changing climates resulted in great migrations of plant and animal communities. Sea-level changes and glaciation controlled important migration routes between the Old World and the New, as well as on the continent itself. Most fascinating of all is the record of massive extinctions of dozens of species of Ice Age animals, leaving important ecological niches vacant. The record also includes the first dim trace of the most important immigrants to the New World: human beings. The last chapter of the fossil record becomes the domain of archaeologists, and the tracking of early people in North America is providing important insights into human influence on the natural history of the continent.

## CHANGES IN VEGETATION

The response of plant communities to the advance of continental glaciers was outlined in Chap. 7. Because southward migrations were not obstructed by high mountain barriers, impassable waterways, or deserts, plant species became reestablished in the new climatic zones without suffering extermination. Such was not the case in Europe, where some temperate species such as sweet gum and magnolia did not survive the earliest glaciation in northwestern Europe because southward migration was blocked by the Alps. No doubt, some North American plant species found local refuges suited to their needs, such as the Driftless Area in the midcontinent, the nonglaciated Yukon Valley of central Alaska, and the mountains that were high enough to stand above the continental glaciers. Fossils from hundreds of localities in North America indicate that, except for changes in distribution, plant genera are little different today than at the beginning of the Ice Age. Few plant extinctions are known, and evolution of new genera can only be surmised because the record is very incomplete.

With the retreat of the Laurentide ice sheet at the close of the Wisconsin Glaciation came a wave of migration that eventually resulted in the present distribution of plant communities. Pollen diagrams from lakes and bogs in the glaciated areas of the midcontinent provide a detailed account of the migration history there. The treeless tundra bordering the ice was invaded initially by spruce forests. Where the ice margin had stagnated, vegetation even became established on the debris-covered dead ice. Kettle lakes, actively forming by the melting of buried ice blocks, received sediment and plant debris from the wet, unstable slopes that surrounded them. This early history is represented in many surviving lakes by a bottom zone of plant "trash" and coarse sediment (Fig. 8-2).

**Figure 8-2** As buried ice in this actively forming kettle lake in the Yukon continues to melt, sediment and plant debris from the collapsing sides are trapped in the bottom of the lake, furnishing a record of the origin and early history of the lake and its surroundings. *(Photo by V. N. Rampton.)*

Continued melting of the ice and warming of the climate caused a northward movement of the spruce forest, which thrives in cool, moist environments. Pine and birch, followed in succession by oak and elm, indicate a change to the temperate conditions that prevail today across the southern Great Lakes states.

Semitropical vegetation became established in the southeastern part of the continent, and a change to more arid conditions resulted in a succession to prairie or desert plant communities in the southwest. The present forest-tundra border in the Canadian Arctic was established just a few thousand years ago. In the last few centuries, the natural vegetation has been so profoundly disturbed by human beings throughout most of the continent that it is almost impossible to discern responses attributable to continuing climatic change.

Along with plant fossils, the lake sediments contain a rich collection of other remains, such as mollusks, diatoms, and freshwater vertebrates, that reflects in a variety of ways the progressive evolution of a lake's ecosystem. Understanding the natural history of such complicated systems allows the prediction of responses to anticipated changes of human origin.

## LAND BRIDGES

A comparison of fossil assemblages reveals that periodic animal migrations took place between the New and Old Worlds. Such exchanges between North America and Asia were made possible during glacial periods because the lower sea level of such times exposed a broad connection in the Bering Strait between Siberia and Alaska (Fig. 8-3). Presently, the maximum water depth

**Figure 8-3** Bering Strait land bridge linked Asia with North America during glacial periods when sea level was lowered. During the Wisconsin Glaciation, the land bridge was at least as wide as the shaded area.

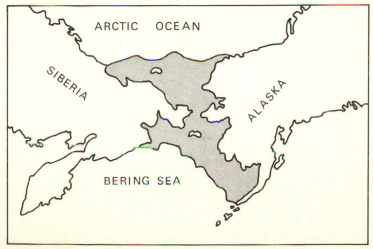

there is about 60 m, and the lowering of the sea by as much as 130 m during the full development of Ice Age glaciers was sufficient to expose not just a "bridge" but a dry land connection that was over 2,000 km wide. Such a broad corridor afforded room enough for the slow diffusion of plants and animals spilled over by population pressure or simply attracted to favorable and unoccupied new territory.

The success of permanent settlement for immigrants into North America depended, of course, upon the synchronous existence of ice-free land in Alaska. Such a refuge is known to have existed in the central interior between the Alaska Range and Brooks Range ice caps (Fig. 6-1). However, southward dispersion into the heartland of the continent was possible only during interglacial periods, or when there existed an ice-free corridor between the Cordilleran and Laurentide ice sheets.

A persistent link between the Americas by the Isthmus of Panama land bridge allowed free exchange between North and South America during the entire Pleistocene Epoch, and even before. That connection accounts for the appearance in North America of such animals as armadillos, ground sloths, and opossums. Eventually that land bridge allowed the invasion of South America by early human big-game hunters.

## ICE AGE ANIMALS

Fossil finds in hundreds of localities throughout North America indicate that the continent supported a wonderful menagerie of animals, many of which became extinct shortly after the ice sheets began to melt away for the last time (Fig. 8-4).

A few of the animals, notably the horse, have long evolutionary lineages well documented in the fossil record of North America and extending back into the Tertiary Period. Others, such as the bison, evolved mainly in Eurasia and immigrated to North America during the Pleistocene Epoch. One notable trend in evolution during the Ice Age was toward large size. Giant-sized beavers, ground sloths, bears, elephants, rhinos, and armadillos abounded. Some evolutionary biologists attribute this increase in size to the cool climate of the Ice Age. Larger animals lose body heat at a slower rate because the larger the animal, the smaller the ratio between its surface area and volume. Other adaptations to cooler temperatures include the development of wool for thermal insulation in such animals as the woolly mammoth and rhinoceros and the growth of thick fat in polar bears as a source of energy in winter.

**Early Faunas**   Fossil localities, mainly in the nonglaciated regions of the Western and Southwestern United States, contain the remains of dozens of different species of animals that lived at the beginning of the Nebraskan Glaciation. The larger mammals include at least three varieties of mastodon, several kinds of saber-toothed cat, and the horse, camel, ground sloth, llama, bear, pronghorn, and jaguar. A great diversity of smaller animals are also represented, many of them species that are living today. These include such common forms as the mouse, rat, fox, skunk, otter, bat, raccoon, shrew, badger, marmot, fisher, mink, and beaver. During the

**Figure 8-4** This woodland scene in the midcontinent depicts a group of Ice Age herbivores, clockwise from top right: mastodont, giant bison, giant ground sloth, and mammoth. Watchful predators include a saber-toothed cat and early humans. *(Drawing by Mary Tanner, technical artist, University of Nebraska State Museum.)*

interglacial following the Nebraskan a number of genera became extinct, notably several kinds of mastodon, at least two types of camel, one saber-toothed cat, and bone-eating dogs.

The middle of the Pleistocene, from the latter part of the Kansan Glaciation through the succeeding Yarmouth Interglacial, is marked by the arrival of a large number of new animals into North America from Eurasia and from South America. Across the Bering Strait land bridge came the first mammoths, several different saber-toothed cats including the famous *Smilodon*, antelopes, musk-ox, and the horse *Equus*. Altogether about eleven Old World genera appear in the fossil record at this time. Most of the genera present at the end of the Kansan Glaciation survived until the end of the Ice Age or are represented in the fauna living today. By the end of the Yarmouth Interglacial, the mastodon *Stegomastodon*, the giant camel *Titanotylopus*, and a zebralike horse had disappeared. Apparently, by this time—halfway into the Ice Age—most of the other animal groups had become adjusted to the fluctuating climates.

**Central Alaska Refuge**   Numerous remains of Ice Age animals, many of them extinct, have been found in frozen deposits along rivers and streams in

**Figure 8-5**   Partial carcass of an extinct bison discovered in frozen silt near Fairbanks, Alaska. It has a radiocarbon age of 31,400 years. *(Photograph no. 600 by T. L. Péwé.)*

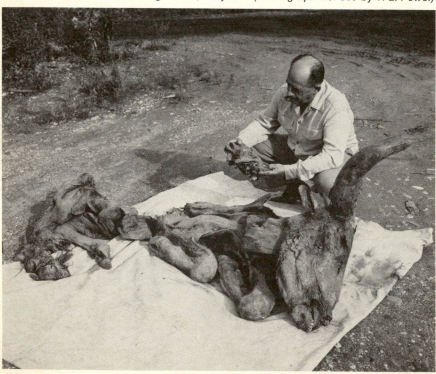

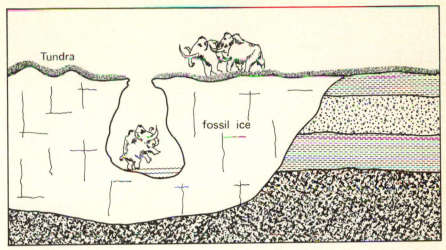

**Figure 8-6** Death and preservation of Ice Age animals might have resulted from the collapse of tundra cover over melting ground ice in permafrost terranes. *(After N. K. Vereshchagin, 1967, Pleistocene Extinctions, Yale University Press.)*

Alaska (Fig. 8-5). Large-scale hydraulic mining for gold in sand and gravel deposits has uncovered hundreds of specimens, including woolly mammoth, ground sloth, mastodon, saber-toothed cat, yak, bison, camel, horse, and musk-ox. In neighboring Siberia, a major ivory trade was based on the exploitation of mammoth and mastodon remains unearthed from the frozen tundra. Preservation of flesh, hair, and skin is common where the deposits are permanently frozen.

Many of the remains appear to have been deposited and buried along river courses during floods. Others were carried into topographic depressions by the slow downslope movements of soil thawed during the Arctic summer. In some instances, the heavier animals may have been trapped by the collapse of the thin tundra surface over melting ground ice (Fig. 8-6). Such accidents would result in a quick death and rapid freezing, resulting in the preservation of the entire animal, including its last meal.

The cold but glacier-free corridor in central Alaska was apparently a refuge for animal immigrants that had crossed the Bering Strait land bridge during glacial periods and a staging area for southward invasions as the continental glaciers melted during the warm interglacials. The Arctic fossil record shows that the main migration was from Asia to America. Just a few "Americans" succeeded in establishing colonies in Asia, notably the reindeer and musk-ox. This predominantly one-way traffic probably reflected the occupation of major ecological niches in Asia, whereas the newly deglaciated areas in North America afforded room for colonization.

**The Rancho La Brea Fauna** The most famous fossil locality in North America is an asphalt deposit in Los Angeles called Rancho La Brea. This

(a)

(b)

**Figure 8-7** La Brea tar pit fauna. *(Mural by C. R. Knight, courtesy of the Los Angeles County Museum.)*

natural petroleum seep acted as a trap for thousands of individual animals during the last half of the Pleistocene Epoch (Fig. 8-7). The asphalt, or brea, is an excellent preservative, and the fossils associated with this deposit are in remarkably good condition, making their study fairly easy. In all, more than 200 different kinds of animals and plants have been recovered from Rancho La Brea. Many of the fossil remains are on view in the Los Angeles County Museum, which stands near the tar pits themselves.

The fauna at Rancho La Brea dates from the Illinoian Glaciation, and the asphalt seeps which have not been covered by the concrete of civilization are still active today. Thus, a remarkably continuous record of life in the Los Angeles region is contained in the deposit. Aside from the sheer number of animals (more than 4,000 individual mammals so far have been excavated), an interesting aspect of the fossil assemblage is the large number of carnivore predators and birds of prey. Apparently the predators and scavengers were lured to their deaths by animals trapped in the tar. Another wave of immigration from the Old World is evident in the fossils of Rancho La Brea. Bison occur for the first time, along with a variety of smaller animals.

The two most common carnivores in the tar deposits are the dire wolf (more than 1,500 individuals) and the saber-toothed cat *Smilodon* (1,000 individuals). Perhaps the most formidable of the predators represented is the great American lionlike cat *Panthera atrox* (Fig. 8-8). Males of this species were nearly one-fourth larger than the modern Indian tiger of Asia. Also present are other predators: lynx, fox, coyote, puma, and a large extinct bear. Herbivores represented include mammoth, mastodon, ground sloth, horse, tapir, camel, and bison. Parts of a human skull and other skeletal remains found in association with these Ice Age animals indicate that human beings were contemporaries of some of these extinct species in Southern California. The bird assemblage consists of 125 different types. By far the most abundant are birds of prey. Such living species as golden eagle, bald eagle, turkey vulture, and a variety of hawks have been found, along with several extinct birds, including a great condorlike vulture that boasted a wingspan of 3.6 m and a body weight of 25 kg. Hundreds of other animal remains, along with many plant fossils, make the La Brea deposits a reliquary of Ice Age fossils, representing a continuous drama of death for both predators and prey for almost 200,000 years.

## EARLY HUMANS IN NORTH AMERICA

The most important mammals to make a successful migration across the Bering Strait bridge during the Ice Age were members of the genus *Homo*. Details of human entry into the New World are lacking because the route has been submerged by the postglacial rise in sea level. Another difficulty in tracing the history of the human race is a lack of appropriate fossil materials.

**Figure 8-8** Reconstructed skeleton of the lionlike cat *Panthera atrox*. This great cat was probably the most formidable predator present in the Rancho La Brea assemblage. *(Courtesy of the Los Angeles County Museum.)*

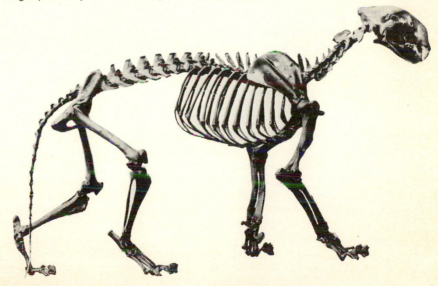

Few actual remains of individuals have been found. The main evidence of their existence and activities is contained in the litter of their campsites. The antiquity of humans is well established in the Old World, and no one doubts that the major evolutionary steps in their development occurred there. Thus, New World archaeology is mainly concerned with establishing the date of entry, tracing migration routes, and deducing the cultural attributes of these early discoverers of America.

The exact time of the first human appearance in America is not yet established with certainty, but the application of $C^{14}$ and other age-dating techniques points to the probability that it was a fairly recent occurrence. The oldest radiocarbon dates from Alaska indicate occupation of land east of Bering Strait by 8,500 years ago; however, sedimentary strata below the dated horizon contain handmade stone tools, evidence of even older habitation. Many radiocarbon dates of kill sites in the Great Plains document the presence of big-game hunters in the midcontinent between 11,000 and 13,000 years ago. Excavations at Clovis and Folsom in New Mexico, and in many other places, reveal strong evidence that two separate groups of hunters preyed on the Ice Age mammals of the Western United States. The earlier group is thought to have been primarily mammoth hunters. Fluted projectile points, called Clovis points, made of bone and flint, along with flint tools for butchering have been found in direct association with mammoth bones (Fig. 8-9). Charcoal from campsites indicates that the earliest of the hunters lived about 11,500 years ago. A more advanced fluted projectile, the

**Figure 8-9** The large tools are typical of Clovis fluted points. The two small ones are Folsom fluted types. *(Photo by K. R. Moran.)*

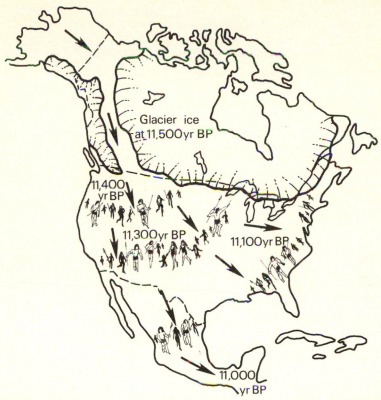

**Figure 8-10**  Model for the migration of early humans into North America at the close of the Great Ice Age. Fronts of denser population are tailed by edges of lesser human density. *(After Paul S. Martin, 1973, Science, vol. 179.)*

Folsom point, has been found throughout North America. At the original site, these projectile points were discovered along with the bones of 23 bison, indicating the killing efficiency of the hunters who fashioned them. Stratigraphic evidence places the Folsom hunters at a relatively later age in America than the Clovis point culture, and $C^{14}$ dating of sites indicates a range of 9,000 to 11,000 years ago for their existence.

A few age estimates hold out the possibility that the first Americans arrived much earlier than the inventors of the Clovis and Folsom projectiles. The skeleton found at Rancho La Brea yielded a radiocarbon date of almost 24,000 years ago, and several other sites indicate ages of greater than 20,000 years. Whatever the timing of the first discovery, it seems clear that an important invasion was launched from the nonglaciated area of central Alaska about 12,000 years ago. Paul S. Martin, of the University of Arizona, has proposed the general pattern of migration as a leading edge of hunters, moving rapidly in pursuit of big game animals (Fig. 8-10).

In just a few thousand years, this wave of human migration had spread across most of the New World, leaving behind a trail of dead Ice Age animals. The cultural evolution of these early immigrants and their successful exploitation of the resources of the New World are richly recorded in the artifacts they left behind.

## EXTINCTIONS

One of the greatest mysteries of the Ice Age is the disappearance of so many of the large animals in the few thousand years following the melting of the continental glaciers at the end of the Wisconsin Glaciation. Most of the animals lost were large mammals, including both herbivores and carnivores. Marine organisms were not affected, nor is extinction evident in the plant kingdom. Although some animals had disappeared during earlier glacial times, as outlined earlier in this chapter, the last wave of extinction was by far the most severe.

Radiocarbon dates of fossil localities indicate that most of the extinctions had been accomplished by 8,000 years ago. By that time such important residents as elephants, camels, horses, and ground sloths were gone. Others that did not survive include the mastodon, giant beaver, long-horn bison, dire wolf, and saber-toothed cat. In all, worldwide, over 200 genera were affected, and neither evolution nor immigration has produced replacements for the lost animals.

What caused this great period of dying? The coincidence of the appearance in North America of skilled hunters with the disappearance of big game animals, such as mammoth, bison, and horse, is irresistible to some scientists. Paul S. Martin has built a particularly strong case for human-caused extinctions. His "prehistoric overkill" hypothesis is based on an apparently close relationship in time between patterns of animal disappearance and the spread of early hunters. Evidence exists that hunting techniques included stampede and fire drive. A mammoth kill site near Dent, Colorado, contained the bones of 11 immature females and 1 adult male, indicating that the young of the hunted species were particularly vulnerable. That human beings are the direct cause of animal extinctions both by hunting and by habitat disturbance in historical times is well documented. The role of prehistoric people in animal extinction might have been similar.

The climatic changes accompanying the end of the Ice Age have been prominently cited as triggering a succession of extinctions. Since most of the genera were large herbivores, a climatically induced restriction of range is considered by some to be the cause of their demise. Increased aridity and loss of grazing land would decimate the larger animals, and in a domino effect, the large carnivores and their dependents would suffer decline as well. An explanation by A. Dreimanis, of the University of Western Ontario, for the extinction of the mastodons is based on direct evidence that these animals preferred a spruce forest habitat. He believes that their extinction is related to the change in vegetation instigated by increased dryness beginning about 11,000 years ago. As the spruce forests were replaced by pine and hardwoods, the mastodons became trapped in wet lowland refuges where spruce persisted. Eventually these habitats changed as well, and the mastodons, unable to migrate to suitable habitat far to the north, disappeared.

By whatever means these Ice Age animals became extinct, their passing, as noted by A. L. Wallace, left this continent and the world zoologically impoverished in "all the hugest, and fiercest, and strangest forms."

# NINE

## THE NATURE OF CLIMATIC CHANGE

Satellite photos gave the first warning. In 1971, seasonal snow cover in the Northern Hemisphere began on Labor Day, almost 2 weeks earlier than usual. Snow accumulated during the winter in record amounts and persisted in snowbanks through late July of the following year. The pattern of early snow and late melting continued. By August 1982, snow and sea ice covered 15 million km², for that time of year an increase of 10 percent in just one decade. That summer, fishing boats remained icebound along the entire coast of Iceland. The grain harvest in the midcontinent was the worst in 50 years. By the eve of the twenty-first century, large permanent snowfields were established in many parts of Canada. Tidal gage readings in Boston Harbor on July 4, 2026, indicated that mean sea level was 10 cm lower than it had been in the bicentennial year of the United States, 1976. The new ice age arrived earlier than climatologists had predicted.

Is such a scenario possible? If the past history of the earth holds any value for predicting future natural events, then the answer is a definite yes. However, paleoclimatologists still lack knowledge in critical areas that would enable the construction of a reliable timetable for future climatic events. The ultimate challenge, still to be successfully met, is the problem of causes of the climatic fluctuations that produced the Great Ice Age. Dozens of hypotheses have been proposed, yet none is capable of explaining all the known characteristics of ice ages. The failure is largely a result of an inadequate understanding of the atmosphere and incomplete documentation of its past history. But each year, as more observations are accumulated, the details of climatic fluctuation, their intensity, duration, and repetition patterns are becoming better known. The time is at hand when the reliable prediction of future climatic trends may be possible.

## ANCIENT ICE AGES

The recent ice age, or in another view, the present ice age, is just one of a series of such intervals of colder climate in the earth's history. From a careful study of ancient rocks, it appears that at least five episodes of prolonged continental glaciation preceded the latest one. As recently as 1961 a field party of French and Algerian geologists, searching for petroleum deposits in the central Sahara, discovered evidence for glaciation 450 million years ago. There were all the familiar signs: scratched and polished bedrock, as well as such features as outwash plains, eskers, and drumlins. The glacial deposits themselves were so old that they had been cemented into a hard rock called *tillite* (Fig. 9-1). Even so, the characteristic features were preserved. The age of the glacial deposits was determined by fossil assemblages and superposition. The glacially abraded bedrock contained remains of organisms of Late Ordovician age, whereas the entire glacial sequence was buried in places beneath marine sedimentary rocks containing fossils associated with the Silurian Period.

**Figure 9-1**  Hammer handle rests on the Dwyka tillite of Permian age. Striations on the underlying bedrock are a sure sign of glacial activity. Harrisdale, South Africa. *(R. F. Flint, Glacial and Quaternary Geology.)*

Evidence for other ancient ice ages has been found in the form of tillite deposits and striated bedrock surfaces on almost every continent. In eastern Canada, the Gowganda tillite is estimated to be in the range of 2.2 billion years old. Glaciation for Russia and China is indicated about 950 million years ago, and Australia apparently experienced ice ages at 750 and 650 million years in the past. In Utah, tillites of about that age are exposed in the Wasatch Mountains. The Dwyka tillite in South Africa is the deposit of a continental ice sheet that existed approximately 300 million years ago.

Thus, an impressive number of glacial episodes have been documented and dated in the geological record. The earlier ones apparently endured for tens of millions of years. Although the recurrence pattern of past ice ages is still imperfectly known, some scientists have suggested a cosmic connection, based on the apparent coincidence of ice ages with certain positions of the solar system in its revolution around the galaxy in a cosmic year, about 275 million of our earth years. Such grand views of earth events may well be valid; however, the lack of precise age dating for many of the glacial events, as well as the probability that not all the episodes have yet been recognized, detracts from their usefulness.

## THE GREAT ICE AGE

Fixing the precise time for the start of the worldwide cooling trend that initiated the Great Ice Age is difficult, but a large amount of evidence indicates that the trend began long before 2 million years ago. The Antarctic ice sheet is at least 10 million years old, according to the geological evidence. Thus, in the broad sense that an ice age is a time when major ice sheets and glaciers have existed on the earth's surface, the birth of the Antarctic ice sheet marks the beginning of the latest glacial cycle. North America and Eurasia developed ice covers much later, and the activities of the Laurentide and Scandinavian ice sheets during the last 2 million years have traditionally provided the basis for an ice-age history. Also, it is this time interval that is best represented in the geologic record, for the deposits, where present, are the uppermost sedimentary units. Therefore, the view that the last ice age has endured for just 2 million years is understandable. The details of climatic fluctuation within this latest epoch of cold are rapidly coming to light, and they furnish a valuable documentation of the patterns of inconstancy in the atmosphere.

Recent oxygen isotope studies of deep-sea cores from the Atlantic Ocean indicate that about 20 warm-cold cycles have occurred during the last 2 million years. Variation in mean earth temperature during these fluctuations is in the range of 6 to 10°C. Each major cycle is roughly 100,000 years long. Numerous minor fluctuations are superimposed on the major trends (Fig. 9-2). Age dating of the cores indicates that a few of the changes from warm to cold were abrupt, the transition having been accomplished in a few hundred years. One such sudden cooling episode at about 90,000 years ago has been detected in the Greenland ice sheet and cave deposits in France from oxygen isotope studies, and in sediments from the Gulf of Mexico from

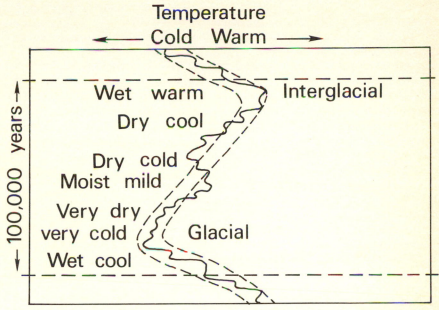

**Figure 9-2**  Climate during a full glacial cycle for the midlatitudes. Notice the many minor fluctuations of short duration that accompany major trends. *(Modified from R. W. Fairbridge, 1972, Quaternary Research, vol. 2.)*

the fossil content. The worldwide nature of these fluctuations is thus fairly well documented.

Most evidence now indicates that the cold part of the cycle is of much longer duration than the warm part. Contrary to early views that interglacial periods were extremely long, the warm periods are thought to endure for just 10,000 years or so. If that is the case, then the present climate of the earth, within the context of the last 2 million years, is abnormal, since interglacial periods make up only about 10 percent of the total length of the last ice age.

Against this background of information from the marine record, the history of continental glaciation must be reconsidered. The traditional view that North America experienced four major glaciations is based mainly on the superposition of tills, with intervening horizons, such as soils, that indicate warmer, nonglacial environments. In Europe, six or seven episodes of glacial activity have been defined from similar observations. Why are there not 20 glaciations, one to match each cold cycle indicated in the marine record? Glacial deposits of many different ages have been discovered in the Rocky Mountains and the Sierra Nevada, some much older than the Nebraskan deposits of the midcontinent. Is it possible that some of the climatic fluctuations expanded glaciers only in mountainous regions? Subsequently, much of the evidence for their existence might have been destroyed by rapid erosion. There is also a chance that the deposits of some continental glaciations have been completely removed by erosion, even by later glaciers. Thus, either the conditions for growth of continental glaciers did not prevail during each climatic fluctuation, or the geological record on the continents has been censored by erosion.

A detailed study of the glacial deposits of the Wisconsin Glaciation, the uppermost glacial deposits, does reveal the complexity of climatic fluctuations during one cycle as they are reflected in the periodic advance and retreat of the Laurentide ice sheet. In the state of Illinois, the Michigan Lobe left an impressive record of moraines and till sheets that indicates at least four distinct fluctuations of the ice margin during the last 70,000 years (Fig. 9-3). Elsewhere, complex stratigraphic sequences and landforms corroborate the unsteady nature of major glacial cycles.

**Figure 9-3** Glacial deposits and intervening nonglacial sediments record the complex fluctuations of climate and glaciers in Illinois and Wisconsin during the last major glaciation. *(Adapted from H. B. Willman and J. C. Frye, 1970, Illinois Geological Survey Bulletin 94.)*

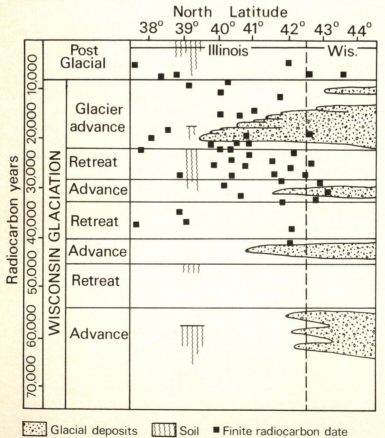

One complication in attempting to correlate continental glacial sequences with the marine record lies in the fact that the growth of a large glacier may take a considerable amount of time. The time lag between onslaught of cold climate and glacial activity, especially of large continental glaciers, may be many thousands of years. Disintegration in response to warming is apparently more rapid.

## THE PRESENT INTERGLACIAL PERIOD

The melting of the Laurentide ice sheet was triggered by an increase in the mean annual temperature of the earth, beginning about 15,000 years ago. The ensuing postglacial period, lasting to the present time, is called the Holocene, or Recent, Epoch. Studies of fossil pollen, tree rings, and the fluctuations of mountain glaciers have resulted in a wealth of data that has been translated into the climatic history of the continent since the great ice sheets retreated. Vegetation patterns indicate a continuous trend toward warmer, more temperate climatic conditions in the midcontinent until about 5,000 years ago. In the fossil pollen record, this climatic warming is reflected in a change from conifer forests to hardwood and then to prairie vegetation. During the next few thousand years, the warming trend was reversed, and the last 5,000 years have been a time of cooler temperatures. The time period between 7,000 and 5,000 years ago is called the *Hypsithermal Interval* because the world temperature was probably about 2 or 3°C higher than it is now.

Since that peak of warmth, the earth has slowly cooled to a condition similar to that which prevailed about 9,000 years ago, only a few thousand years after the retreat of the ice sheets. The change to a cooler, moister climate favored the growth of mountain glaciers in western North America, and this time of renewed glacial activity is known as the *Neoglaciation* (Fig. 9-4). Careful studies of glacial deposits along the margins of present-day glaciers reveal that in the last 6,000 years, glacial expansion occurred during three distinct intervals of time (Fig. 9-5). The last of these, which began in the fourteenth century, is marked by such impressive glacial activity that it is called the Little Ice Age.

Historical records from the European Alps indicate that the Little Ice Age is composed of several episodes of advance and retreat of mountain glaciers. Villages in the Swiss Alps were overrun by ice in the early seventeenth century. In Norway, glaciers advanced as much as 2 km between 1700 and 1759, and the climatic cooling that accompanied the expansion resulted in widespread crop failures and famine. Farms built near glacier margins in Iceland prior to A.D. 1200 were destroyed by advancing ice early in the eighteenth century. A reversal in the trend came in the middle of the nineteenth century, and glaciers began to retreat all over the world. A cooler trend beginning in about 1945 has resulted in the beginning of another episode of advance.

All these observations reveal how complex the climatic patterns really are. Large cycles of cold-warm have intermediate fluctuations superimposed on them, and these in turn are composed of even shorter variations, in a

**Figure 9-4** Neoglacial moraines along the terminus of the Donjek glacier, Yukon Territory, are just 200 to 300 years old. They represent the maximum position of the glacier during the past 10,000 years. *(Photo by Austin S. Post, United States Geological Survey.)*

hierarchy that is remindful of a situation pointed out by Jonathan Swift in "On Poetry, a Rhapsody":

So, naturalists observe, a flea
Has smaller fleas that on him prey;
And these have smaller still to bite 'em;
And so proceed *ad infinitum*.

A comparison of the present interglacial with those of the past indicates that the Holocene is in its waning stages as a warm interlude between ice ages. Ten thousand years or more have already passed, and this seems to be

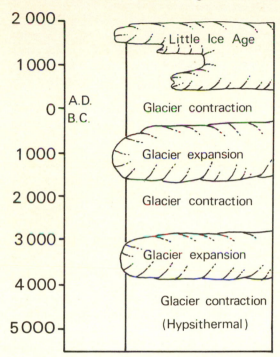

Worldwide Holocene glacier fluctuations

**Figure 9-5** Records of glacier fluctuations during historical times, indicating three intervals of glacier expansion during the last 6,000 years. *(Modified from G. H. Denton and W. Karlén, 1973, Quaternary Research, vol. 3.)*

the average duration of interglacial climates. Thus, the next ice age may be at hand (and with the world in the midst of an energy crisis!).

## CAUSES OF ICE AGES

The search for a cause of ice ages has been pursued vigorously ever since Louis Agassiz's provocative outline of the glacial theory nearly 150 years ago. Two basic events in the climatic history of the earth need to be explained: (1) long-term periods of worldwide glaciation occurring every 200 to 300 million years and (2) climatic fluctuations within the long-term ice ages, resulting in the periodic growth of continental glaciers, with a duration of about 100,000 years. A successful theory must explain the mechanism for long-term changes of the earth's temperature and the periodicity of those fluctuations. The following summary of ideas is a testament to imaginative scientific inquiry, and at the same time it is an admission that the problem of causes has not yet been solved.

**Solar Variability**  Since the sun is the source of the energy that warms the earth's surface, variations in output of solar radiation would be directly reflected in temperature changes in the atmosphere. A decrease of 10 percent in the intensity of sunlight in space could account for the onslaught of an ice age. The well-known 11-year sunspot cycle involves fluctuations of solar output of up to 4 percent. Unfortunately, little is known about changes in solar activity over longer periods of time. At present, this "flickering sun" hypothesis, although attractive because of its simplicity, must remain just a speculation until proof of long-term changes in the sun's output of energy is at hand.

**The Astronomical Theory**    Irregularities in the earth's orbit and its axis of rotation result in three cyclic disturbances in planetary movements that influence the distribution of solar heat on the surface: (1) The shape of the earth's orbit varies between a nearly perfect circle and an ellipse. The time period between maximum change, or eccentricity, is about 90,000 years. (2) The angle of tilt of the earth's rotational axis varies as much as 3°, ranging from 21.5 to 24.5° over a period of about 41,000 years. (3) The direction of the axis in space changes, like a great inclined spinning top. The period of one wobble, or complete precession, is about 21,000 years.

A Yugoslavian geophysicist, M. Milankovitch, concluded that the joint effect of these three orbital factors was an alternation of long periods of cool summers with periods of warm summers in the northern latitudes. Although these oscillations have been confirmed as real, some scientists consider that the changes are too small to cause an ice age. Others believe that these variations act as a triggering mechanism and that subsequent events, such as the growth of glaciers themselves, would account for the remainder of the temperature change. This approach does not account for the long-term occurrence of ice ages, nor does it explain the timing of all the glacial events during the Pleistocene Epoch.

**Changes in the Composition of the Atmosphere**    The composition of the atmosphere affects the planetary heat budget in several ways. Turbidity from particles of dust in the upper air results in a backscattering of solar radiation. It has been suggested that periods of explosive volcanic activity in the earth's history trigger glacial climates because the contamination of the atmosphere with large quantities of volcanic dust would significantly reduce the amount of sunlight reaching the earth's surface. Sharp drops in world temperature have been noted to follow eruptions in historical times. However, these cooler periods have endured for just a few years, because of the short residence time for dust in the atmosphere. Detailed information on the timing and duration of past volcanic activity is inadequate at present to correlate with the major climatic changes associated with ice ages.

Once the sun's energy reaches the earth's surface, carbon dioxide ($CO_2$) in the atmosphere prevents heat from escaping back into space. This "greenhouse effect" maintains an average earth temperature much higher than it would be otherwise. Decreasing the amount of $CO_2$ in the atmosphere would be like opening the windows of a greenhouse—heat would escape more rapidly and the temperature would drop. The $CO_2$ content of the atmosphere is dependent upon a number of complicated processes, including rates of volcanism, weathering of rocks, and storage capacity of dissolved $CO_2$ in the oceans. Our incomplete understanding of all the interactions of these various complex systems makes identification of a specific sequence of cause and effect difficult.

Human activities have resulted in measurable changes in the composition of the atmosphere, especially since the Industrial Revolution. The burning of fossil fuels has introduced more $CO_2$ and thus a tendency toward warmer temperatures. But the pollution of the atmosphere with particulate matter has had the opposite effect. Continued contamination of the air may eventually lead to a human-caused climatic change of significant proportion. Whether the shift is toward a warmer or colder condition depends upon the effects of the various pollutants in relation to the earth's heat budget.

**Geographical Theories**   Various theories have been constructed based upon the relative positions of land and sea and the height of continents above sea level. The general requirement for an ice age in most of these explanations is the location of a major landmass near one of the poles. High elevations would enhance the development of glaciers, and a source of moisture from nearby oceans would nourish them. The expansion of glaciers would increase the amount of solar radiation reflected back into space, thereby cooling the entire earth, because white glacier surfaces are better reflectors than darker land surfaces. Lower snow lines would result, and glaciers could form wherever the land surface was high enough and snowfall sufficient.

During the past decade, quite conclusive evidence has been discovered indicating that continents have moved about on the surface of the earth throughout geologic time. This theory of continental drift and sea-floor spreading allows the determination of the positions of landmasses at various times in the past. Such studies indicate that the present pole positions were attained early in the Cenozoic Era. The fact that glaciation did not begin until the latter part of the Cenozoic flaws this view of cause. Although the positions of Antarctica and Greenland today account for the ice sheets they support, none of the geographical hypotheses adequately explains the glacial-interglacial fluctuations that are characteristic of ice ages.

**Surging of the Antarctic Ice Sheet**   An interesting hypothesis being investigated at the present time ascribes the periodic glaciations of the Northern Hemisphere to the activity of the Antarctic ice sheet. In this model for ice ages, proposed by A. T. Wilson, of Victoria University, Wellington, the growth of that ice sheet is out of step with the Northern Hemisphere glaciers. During interglacials such as the present, the Antarctic ice thickens, as it is fed by moisture evaporated from the Antarctic Ocean. Heat from the earth's interior warms the base of the ice to the melting point, and the ice sheet begins to slide outward. The result is a rapid expansion of an ice shelf around the continent that increases the amount of solar radiation reflected away from the earth. The ensuing temperature drop initiates growth of glaciers in the Northern Hemisphere, further reducing the temperature of the earth. The thinned Antarctic ice sheet becomes colder, and rapid sliding ceases when the basal ice refreezes to the bedrock surface. Climatic warming begins when the starved ice shelf breaks up by rapid calving, decreasing the reflecting capacity of the earth. This signals the end of the glacial period, and northern ice sheets begin to melt. Eventually the chain of events repeats itself.

We shall have to wait for geological evaluation of this interesting hypothesis. Especially crucial will be evidence in the deep-sea sediments of the Antarctic Ocean that ice shelves have been extensively and periodically developed in the past. The surging of ice into the Antarctic Ocean would produce a rise in sea level just prior to the general deterioration of world climates. Evidence for such changes is not yet at hand.

**Magnetic Field**   Recent studies of deep-sea cores by Goesta Wollin and his colleagues at Lamont-Doherty Geological Observatory show an apparent link between trends of climatic change and variations in the earth's magnetic field through time (Fig. 9-6). High intensity of the magnetic field appears to correspond to periods of colder climate. It is therefore possible that the earth's magnetism may have a shielding effect against solar radiation, thus modulating the earth's climate. If this is so, then long-term climatic changes may be related to the activity of the earth's core, which is believed to be the source of the magnetic field. Other observations, however, indicate that solar

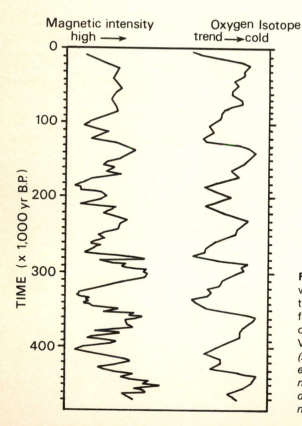

**Figure 9-6** Correlation of variations in magnetic intensity and a climate curve derived from oxygen isotope studies of deep-sea sediment core V12-122 from the Caribbean. *(After Goesta Wollin and others, 1974, Colloques Internationaux du Centre National de la Recherche Scientifique, no. 219.)*

activity directly instigates changes in the magnetic field, thus supporting the possibility that both climate and small changes in the magnetic field are controlled by processes in the sun.

## CONCLUSION

In a speech dedicating a new Quaternary Research Center in 1973, R. F. Flint, of Yale University, summarized the great scientific theories that have unified our views of the activities that take place at or near the surface of the earth. In the nineteenth century came the theory of organic evolution, stated so ably by Charles Darwin. That theory explains the mechanism by which the biosphere has constantly changed through time. The puzzle of the fossils and the origin of species was solved. An understanding of the dynamics of the outer shell of the earth, the lithosphere, came in the twentieth century. In fact, the full impact of discovery is still fresh upon us, as the theory continues to be tested. Within this theory of the dynamics of the lithosphere, the origin of ocean basins is explained as well as mountain building and continental drift. The third great physical system, the atmosphere, still eludes the formulation of a unified theory to explain its activity through time, perhaps because it is the most complex system of all. But such a theory of climatic variation may not be far off. Professor Flint, world-renowned for his research on ice ages, concluded his dedication address with the following words: "Our strong curiosity, our increasing manpower, and the rapid development of our technical skills lead me to expect that the theory of climatic variation will be with us before the end of the 20th Century."

Thus, the puzzle of the Great Ice Age may soon be solved, and with the solution hopefully will come the understanding necessary to help ensure the continued existence of human life on this most extraordinary planet.

# SELECTED READINGS

**CHAPTER 1**

Agassiz, Louis, *Studies on Glaciers*, Albert V. Carozzi, ed. & tr., 1967, Hafner Publishing Company, Inc., New York, 213 p.

Chorley, R. J., Dunn, A. J., and Beckinsale, R. P., 1964, *The History of the Study of Landforms*, Methuen-Wiley, London and New York, 678 p.


**CHAPTER 2**

Andrews, J. T., 1975, *Glacial Systems*, Duxbury Press, Belmont, Calif., 191 p.

Dyson, J. L., 1962, *The World of Ice*, Alfred A. Knopf, Inc., New York, 292 p.

Paterson, W. S. B., 1969, *The Physics of Glaciers*, Pergamon Press, New York, 250 p.

Post, A., and LaChapelle, E. R., 1971, *Glacier Ice*, University of Washington Press, Seattle, 110 p.

Sharp, R. P., 1960, *Glaciers*, University of Oregon Press, Eugene, 78 p.


**CHAPTER 3**

Coates, D. R., ed., 1974, *Glacial Geomorphology*, Publications in Geomorphology, State University of New York, Binghamton, 398 p.

Flint, R. F., 1971, *Glacial and Quaternary Geology*, John Wiley & Sons, Inc., New York, 892 p.

Price, R. J., 1973, *Glacial and Fluvioglacial Landforms*, Hafner Publishing Company, Inc., New York, 242 p.


**CHAPTER 4**

Eicher, D. L., 1968, *Geologic Time*, Prentice-Hall, Inc., Englewood Cliffs, N.J., 149 p.

West, R. G., 1968, *Pleistocene Geology and Biology*, Longmans, Green & Co., Ltd., London, 377 p.

Zeuner, F. E., 1970, *Dating the Past, An Introduction to Geochronology*, Hafner Publishing Company, Inc., New York, 516 p.


**CHAPTER 5**

Ladurie, L. E., 1971, *Times of Feast, Times of Famine*, Doubleday & Company, Garden City, N.Y., 426 p.

Schwarzbach, M., 1963, *Climates of the Past*, D. Van Nostrand Company, Inc., Princeton, N.J., 328 p.

Turekian, K., ed., 1971, *Late Cenozoic Glacial Ages*, Yale University Press, New Haven, Conn., 606 p.


**CHAPTERS 6 AND 7**

Flint, R. F., 1971, *Glacial and Quaternary Geology*, John Wiley & Sons, Inc., New York, 892 p.

Wright, H. E., Jr., and Frey, D. G., eds., 1965, *The Quaternary of the United States*, Princeton University Press, Princeton, N.J., 922 p.


**CHAPTER 8**

MacNeish, R. S., ed., 1972, *Early Man in America*, Readings from Scientific American, W. H. Freeman and Company, San Francisco, 93 p.

Martin, P. S., and Wright, H. E., eds., 1967, *Pleistocene Extinctions: The Search for a Cause*, Yale University Press, New Haven, Conn., 453 p.

Stock, C., 1963, *Rancho La Brea*, Science Series No. 20, Los Angeles County Museum, 83 p.


**CHAPTER 9**

Mitchell, J. M., ed., 1968, *Causes of Climatic Change*, American Meteorological Society, Boston, 159 p.

# INDEX

Page numbers in *italic* indicate pages on which there are photographs or figures.

# MATT MILLER IN THE COLONIES

*Book One: Journeyman*

MARK J. ROSE

*The Skydenn Looking Glass*
*Simi Valley, California*

The Skydenn Looking Glass
508 Longbranch Rd.
Simi Valley, CA 93065

Printed in the United States of America

Library of Congress Cataloging-in-Publication Data
Rose, Mark J., 1965 –
Matt Miller in the Colonies : Book One: Journeyman / Mark J. Rose.
ISBN 978-0-9975554-1-7
1. Science Fiction. 2. Historical Fiction.
Title: Matt Miller in the Colonies : Book One: Journeyman

Second Edition

# CHAPTER 1.

## TOOTHPASTE, PART I

The physicists took care not to step on the eight legs of the hissing stainless steel spider as they worked to adjust the power supplies that formed the magnetic containment field. None of the four was even remotely aware that they adjusted a machine that had the power to change history. Brian Palmer, a stocky man with an unruly mop of dark hair and a mustache, used a rag to clean the glass portal on top of the reactor that would allow them to observe the Chernenko-Einstein particles as they formed. Although they had an overwhelming trust in the analytical equipment and sensors that would measure the reaction, there was no substitute for seeing it with their own eyes.

Jacob Cromwell, a tall, thin scientist with red hair, spoke with a thick Southern drawl. "What time is it?"

Palmer set the rag down and checked the phone in his pocket. "Six a.m.," he replied.

"Seriously?"

"Yes," Palmer said, discouraged. "I can try one more and then I gotta go. I have a presentation to give in four hours. I'll be sleepwalking through it."

"Fine by me," Kevin Moore, the third physicist, answered. "My hands are starting to shake from all the coffee." Kevin then turned to a fourth man who had worked his way under the reactor holding a crescent wrench. "David, you good for one more?"

"Might as well," David Greer replied as he slid partially out from underneath the stainless steel monster. "It's a wasted day for me either way. I can't function on three hours of sleep."

"One more try, then," Palmer declared. "Almost ready?"

Greer reached up while still on his back to make a last-minute change. "Done, I think," he said. He slowly took his feet and backed away from the reactor. Each stood close enough to see into the glass portal.

"Wait," Palmer said calmly. They watched as he tried to squeeze between the reactor and the wall. Palmer reached into his pocket, pulled out his phone and set it on the shelf. The empty pocket gave him just enough space to squeeze through and turn the last calibration knob. He eased his body out and joined the rest to watch the product of two hours of adjustment.

Cromwell pulled the switch. A steady vibration could be heard from the reactor and then a green glow started to emanate through the glass. The men looked on with some interest. After so many attempts, all were surprised to see the reactor finally begin to produce particles.

**********

Outside, the town of Oak Ridge was beginning to wake. The residents were unaware of the experiment that was taking place in the basement of the government laboratory, but four people would soon experience the direct effect of a concentrated beam of Chernenko-Einstein particles as they escaped containment.

One of these four, a mountain biker named Patrick Ferguson, was starting to think that fall would never end in Tennessee. October had been unseasonably warm, resulting in the most brilliant colors the state had seen in decades. He was making a habit of waking early to be in the wooded hills as the sun rose to light the leaves. Oak Ridge was a stunningly dull place to live compared to his hometown of London, but for sheer beauty, it would've been hard to beat the southern United States in the fall. He was breathing hard as he pedaled his bike through the countryside and listened to music from the phone in his breast pocket. He felt the phone vibrate, but it was unlike the vibration it made for a call. He reached for his pocket to verify that the phone hadn't fallen out.

A teenage girl, Sarah Morris, was holding her phone when the reactor started, so the sensation felt more like an electric shock. She studied the phone, wondering if something was wrong, but her mother driving up in a black Mercedes-Benz interrupted her thoughts. By the time the car stopped, she had entirely forgotten about the phone. Her mother got out of the car in the long black dress and short grey fur she had worn to a charity fundraiser the night before. She reached down to pick up her daughter's violin case.

"How was it?" Sarah asked.

Her mother gave a noncommittal nod. "I donated more of your father's money," she said.

"How'd you sleep?"

"You know I hate hotels, but it was too late to drive home."

The last person to experience the effect of the Chernenko-Einstein experiment was a twenty-six-year-old
hiker named Matt Miller. He was high up in the Smoky
Mountains, trying to interpret a paper map using the
compass function on his phone. When the shock came,
it made his fingers go numb and the map drop from his
hand. Irritated, he glared at his expensive new phone.

The only other creature aware that the reactor had successfully begun to generate a Chernenko-Einstein particle string was a squirrel that was sitting on the Rutherford
Q-band cell phone tower high atop Clingmans Dome,
the highest peak in the mountains of Tennessee. The 'Q'
stood for quantum, which was the new cell technology
that had been introduced recently by the Rutherford
Company. Of all the creatures feeling the effects of the
successful generation of Chernenko-Einstein particles,
the squirrel was in the best position to realize that the
sensation they felt wasn't an electrical shock at all, but
that of matter being disrupted by the first few particles to
escape the reactor.

**********

In the lab, three of the four physicists stared intensely at
the humming metal bug. Cromwell kept his eyes glued to
a computer monitor as he called out readings. "Ten thousand...twenty thousand...thirty thousand...man...it's
fifty thousand!" He glanced away to look through the
reactor window. The room had begun to take on a green
hue from the particles interacting with the magnetic field.
The faint green light was accompanied by a calm hum.
Cromwell looked back to the energy monitor and called
out with excitement, "Holding at one hundred and twenty
thousand!"

"My God," Palmer exclaimed, "we've done it!"

They looked at each other, smiling. "You have data?" asked Greer.

"Tons," replied Cromwell. A bright green flash and then a sharp snap like the sound of a bullwhip punctuated the end of his sentence. Their gazes were frantic. The only thing more startling than the fact that the reactor had begun working was that it had as abruptly stopped. Cromwell scrutinized the monitor. "They're gone," he said.

"Gone where?" Greer asked. "The trap wasn't open." He watched as Cromwell flipped switches to the off position.

"The containment field's broken," Palmer said.

"Where'd they go?" Greer repeated.

Palmer stooped down to inspect the reactor casing. "Probably through here," he said as he moved his two fingers inside a round hole in the stainless steel housing. "Cold and smooth, like it disappeared." With his eyes, he followed the angle of his fingers to where they pointed at the wall. A hole had been bored through the laboratory shelf. He stood to retrieve his phone and found that a perfect two-inch core was missing from the screen. They all inspected the damage as he turned it in his hand. "This cost me half a month's salary." All their eyes returned to the shelf.

"They went through the wall!" Greer exclaimed. The physicists followed him into the next lab. The hole was there too, but this time bigger, exiting through the ceiling. They ran together up the steps to the first floor. Cromwell waved his badge to open the secured door and glanced into the lab directly above the reactor. He pointed down the hall. "Entrance and exit holes going straight that way," he said. "East side of the building."

They hurried out into the morning light to inspect the outside. "They didn't make it out," Moore said, relieved. He was watching Greer rounding the corner.

"Yes, they did," yelled Greer.

"They would've had to change direction!" Cromwell said doubtfully, rushing to join Greer. There was a hole through the bricks in the southeast corner of the building. The men arrived in time to watch a large oak tree begin to teeter where a two-foot section had disappeared from its trunk. The tree crashed to the ground, sending bright fall colors and dust in every direction. Once the tree was gone they saw the circular path that the particles had carved through the forest on their way to the town of Oak Ridge. They stood paralyzed by the destruction.

"We better call Colonel Gabriel," Cromwell said.

# CHAPTER 2.

# MATT MILLER

Matt awoke with the violent gasp of a drowning man. Aside from this, nothing seemed out of the ordinary as he lay there on a makeshift bed looking up into the wooden rafters of a large barn. He had slept in many hiking shelters in the last few years and was used to waking up under rustic rooftops. He breathed in deeply. It was a bouquet of hay, dirt, and manure; the wholesome smells of a farm. Matt stretched his arms and legs to rise from the bed but his sore muscles fought him. He relaxed, trying to remember what he did the day before that would explain his aching limbs. His body hurt more than it should for the third week of a hike.

His mouth was dry, so he gently rolled over on the wooden bench to grab water from the pack underneath. He felt around, then reached again deeper, but still grasped only air. He dropped his head over the side of the bench but it was too dark to see. Light streaming through the barn's glassless windows threw long shadows and made the things around him barely visible, so he eased himself onto the hay-covered floor for a better view. The space below him was empty and his pack was nowhere to be found. Someone had walked off with his stuff. *The gun!*

His mind raced. Matt had bought the pistol one day after a mama grizzly chased him up a tree. He'd been uncomfortable carrying it right from the start, so instead of strapping it at his side, where he could actually use it against an attacking bear, he kept the pistol packed away in his bag, unloaded. The box of bullets was tightly wrapped because he had the suspicion that they would be jarred somehow and start going off like firecrackers. The gun, along with everything else he carried, was expensive, so he imagined there was a happy thief out there somewhere.

Matt stood upright and a searing pain went through his head, causing his knees to buckle. This was more than soreness from the trail. He pulled himself back onto the bench to catch his breath and wait for his headache to diminish to a dull throb, then eased himself onto his feet. He fought off dizziness as he walked to the barn door and slid it open. The light was blinding and the pain hit him again, but he managed to remain standing. He squinted to see a farmhouse in the distance, framed by large oak trees. Chickens clucked to his right, and the wind was rustling the leaves. He stood there, trying first to steady his legs, and then to get his bearings. He had expected to walk out into something familiar, but nothing was recognizable. Nothing looked at all like the Appalachian Trail.

"Good afternoon to you, sir."

Matt turned toward the chickens, shielding his eyes from the sun. Standing behind a wire coop was a tall middle-aged man in a worn white shirt and faded blue trousers. He was using a hammer on a coop fence as chickens frantically pecked at and around his feet. Matt struggled to say something. The trail often brought him face to face with poor mountain people, and this was no exception. He wasn't always comfortable around these people. These thoughts and the pain in his head clogged his response, but he finally managed to say hello.

"We thought you'd never awaken," the man called.

Seemingly from nowhere, a large shepherd dog trotted forward. Matt stepped back as the dog stood to face him and barked loudly. "Easy, Cujo!" Matt said. The dog stopped as soon as he spoke and tilted his head as if he understood, but after a moment resumed barking.

"Scout!" the man said. "Come hither!" The dog looked to the man and then grudgingly trotted to the coop. He split his gaze between Matt and his destination for the ten yards it took to go to his master, then sat down next to him and continued to growl. "Calm down, dog," the man said as he stepped between the chickens. He walked to Matt and reached his arm out. "Thomas Taylor," he said.

"Matt Miller," replied Matt, reaching his arm out to shake. "How long have I been sleeping?"

"Almost two days."

"Two days?" Matt said, surprised. "How did I get here?"

"We fetched you from the road," Thomas replied.

"The road?" Matt looked around, head still pounding, trying unsuccessfully to remember how he got into the barn. Finally he thought to ask, "Have you seen a backpack?"

"In the house," Thomas replied. "I'll wager you're not a Frenchman." He grinned.

"Why would I be French?"

"My sons thought this by your clothes."

It made sense to Matt. The Appalachian Trail attracted thousands of people every year, many from Europe. "No, I'm American," Matt replied. "Could I get my pack?"

Thomas nodded and said, "I'll be finished in a moment."

"Can I use your bathroom?" Matt asked.

"A bath?"

"Not a bath," Matt said, motioning downward to the center of his body. "Somewhere to pee." Thomas looked at him strangely until he noticed the hand motion, and then he pointed to a narrow outhouse a good distance from the barn. The dog started to follow as Matt walked away, but Thomas called him back sharply.

By the time Matt returned, Thomas was finished with his fence. "You should have a meal before you leave," he said. This man seemed too well-spoken to be living in poverty.

They walked to the house and the dog followed immediately behind them, making half growls to let Matt know he was still there. Matt climbed the six steps onto the porch, steadying himself against the railing. He didn't know what to expect as he entered the home and was surprised that it was clean inside. He didn't see a television, or anything else that required electricity. Matt scanned the floor for electrical outlets and saw none. *Do people still live like this?*

There was a kitchen at a lower level directly in the back of the house. A woman was working in front of a soot-stained hearth. Matt could smell the burning wood. As he walked into the house, there was no longer anything to hold to steady himself, and he became disoriented. The room grew rapidly hot, and then a cold feeling washed over his body and his vision blurred. He couldn't stop himself as his legs gave out and he collapsed to the floor unconscious.

*********

Matt woke in the shade of the oak trees behind the house, looking into the kitchen. He pulled himself upright in a wooden chair next to a table. "Drink this," a woman standing over him said with a German accent as she held out a tin cup of water. She was dressed in a simple country dress and her hair was pulled back. Her teeth were straight but yellowed and Matt became obsessed with her mouth for more than a moment, caught himself staring, and looked away. Obviously, she hadn't been to a dentist in a while.

"You need something in your stomach," she said. "I'll bring you victuals." She returned with a tin plate of beef stew, a cooked cob of corn and a hard piece of bread, and then left to go back into the house. When she finally returned, she sat across from Matt and was immediately pleased with his empty plate. "Seems even Frenchmen must eat," she said. She had been right; filling his stomach had worked to bring him out of his daze.

"My name's Matt Miller," Matt said. "I'm American."

"And I'm Mary Taylor," she replied.

"Not every hiker on the trail is from Europe," Matt said, chuckling. She looked at him, puzzled. Matt noticed her reaction but decided to remain quiet. It didn't seem worthwhile to try to explain his situation to this random woman. "Can I get my pack?" he asked. "I probably should be moving on."

"You're ready to return to the road?" she asked, concerned. "A moment ago you couldn't stand." She got up before he could answer and returned struggling with the pack on her shoulder. This surprised him. He had suspected it might be hard to collect his things and be able to leave, but now, here he was, ready to go. She hefted it to him. He zipped it open and took inventory, feeling down to the bottom. As far as he could see, nothing was missing. The gun was in the bag along with the bullets.

Matt's paranoia wasn't lost on the woman. "'Tis all there. We are Christian people," she said. "My husband has your knife. You can have that when you go." She said this like it needed no additional explanation. "Where will you travel?"

"Probably back on the trail," he answered.

"Which trail?"

"The Appalachian Trail."

She seemed puzzled again at the mention of the trail, and then said, "I have a meal to tend." She turned and stepped back into the kitchen, out of sight.

Matt reached into his pocket, pulled out his cell phone, and looked for bars. "I thought you were supposed to get service everywhere," he said, irritated. He'd have to ask directions. Matt stood up with his phone in his hand, still looking for a signal. He felt better now and finally confident that he wouldn't collapse. He pushed his pack behind the kitchen door and slowly walked around to the front of the house, holding the phone high. When he finally looked at his surroundings, he was surprised to see the farm was bigger than he had first observed. The barn where he had slept was directly ahead. There was a sizable horse barn to its left with a number of curious animals sticking their heads out from their stalls.

His first impression had been that these were poor mountain people. He was now rethinking that assumption. Poor mountain people didn't have European accents and a barn full of horses. He stuck his phone in his pocket, wandered to the barn and began to pet a towering chestnut-colored animal. The horse looked down at him through calm brown eyes, shifting his head as Matt scratched, clearly enjoying the attention.

"His name's Thunder," a woman said over his shoulder.

Matt turned to see a young woman brushing a silver horse in another stall. She had pulled-back blond hair and was dressed in a long plain blue and white dress. Matt had to look away to keep from staring; she was a radiant Nordic beauty with ice-blue eyes that hit him like a thunderbolt when he met her gaze.

"Do you ride well?" she asked.

"I don't know anything about horses," Matt replied, laughing. He followed the motion of her hand as she moved the brush with long strokes.

"It must take ages to travel anywhere, then," she chided.

"I do okay," Matt said. He had never, in his life, wanted to learn to ride a horse.

"You may be able to find work," she said. "It's almost hay time and Father will be taking workers." She went back to her task.

"I have a job," Matt said. "How many horses do you own?"

"About twenty grown, ready to sell, and many younger ones," she replied. "We are selling Patriot here. Would you purchase him?" There was a hint of mirth in her voice, as if she assumed he couldn't afford the horse.

"I may," Matt replied. He caught her smirking. "Who are you?"

"Grace Taylor."

"Mary's daughter?"

She nodded. "Thomas Taylor is my father."

"Who owns this farm?"

"Father paid off the debt when I was a little girl."

"Why do you live like this if you have all these horses?"

She looked at him, irritated. "It's true that we may not live as well as some in town. We give much to maintain the church and help the poor. We choose not to live with English luxuries."

"What, like Range Rovers?"

She stared at him as she brushed, but didn't reply.

"My name's Matt Miller," he said, offering his hand. She continued brushing, and he saw her demeanor change.

"Imagine, a drunkard sleeping in our own barn," she whispered towards the horse, loud enough for Matt to hear.

Matt looked at her questioningly. "You think I'm a drunkard?"

"That's how they found you," she said.

"Found me drunk?"

"Yes, lying under Bonner Bridge."

"Bonner Bridge?" He laughed, confident in the fact that he had not slept under any bridges.

"Bonner Bridge," she retorted.

"I haven't had a drink in weeks," he exclaimed.

"You do drink rum, then?" It was both an accusation and a question.

"I'm not a drunkard!"

A young boy appeared at the entrance to the barn, interrupting them. He was quiet.

"What is it, Jonathan?" Grace said.

He turned to Matt. "Father wants to know if you desire work."

"Tell him thanks, but I have a job," Matt replied politely.

"Fine," the boy said. He lingered for a moment and then was gone.

Grace said sharply, "I must finish my chores."

"I'm going," Matt replied. He turned to leave, trying his best not to give her the dirty look he felt she deserved. He walked back to the house, rationalizing his irritation. It was probably because her appraisal was closer to the truth than he wanted to admit. It was true that he hadn't had a drink since the night before he'd left on his trip, but there had been plenty of alcohol before that.

In the past, he had gone hiking to give himself a mental break from his job at the lab. This time, though, the trip had been a physical necessity. Matt had forced himself to leave Philadelphia to escape a routine of partying and drinking that threatened to overwhelm him. This was more than a hiking trip; it was a three-week exile from chemical destruction. It had taken him a week of walking to begin feeling like himself again. By the end of his second week, his pants fit properly and he was no longer craving alcohol.

Matt had planned to use the third week to sort out his future and his relationship with his girlfriend, Kylie. She was his ticket to the beautiful people in Philadelphia. Stunning in her glittering cocktail dresses, she had shown him eight months of the best houses, clothes, and liquor that money could buy. Her parents bankrolled her social life and she lived with them in a sprawling downtown apartment. They had only been home for a few days in the time that Matt had known Kylie; they spent most of their time in Europe or the Mediterranean.

Matt was something of a curiosity among Kylie's friends, who all seemed to be living off some sort of trust fund. It was hard for him to hide his working-class background, or the fact that he had a day job, as he struggled to keep some semblance of balance between his social life and his career. The friends had stopped offering him cocaine, which like all drugs he avoided, but he drank as much as the rest of them. His drinking, along with his beautiful girlfriend, seemed to be enough to keep him an accepted member of their club. Lately, though, Matt had started to wonder if he could continue to pay the price. He felt like his life was slowly being drained away.

Matt's thoughts were interrupted as he reached the house. He headed around back to retrieve his pack from behind the kitchen door. He'd planned to grab it and start on his way, but weakness overwhelmed him and his head began to pound. He sat down in one of the wooden chairs on the porch and poured a cup of water from the pitcher Mary had left. The fact that no one was around was vaguely puzzling after what had seemed like a lot of attention. Could it be that he wasn't the center of the world for these mountain people? He leaned back in the chair and looked up at the wispy white clouds as they floated across the blue sky, silhouetting the green trees.

*The trees are green!* The last time he could remember, it was the middle of fall.

# CHAPTER 3.

# TOOTHPASTE, PART II

---

Ordinarily after such a high-profile accident, Senior Counsel Jane Schaefer would be mopping the floor with the four men around the table. She thought better of it, though, with Colonel Alan Gabriel sitting across from her. She had pulled his dossier before she left her office at the NSA, and between these documents and her own network she learned that he was respected by his peers and connected to the highest echelons of the US government. Colonel Gabriel, now forty-eight, had been in the Marine Corps since enlisting at the age of seventeen. As a young man he did three tours of duty in the Persian Gulf and was twice decorated for valor. The Marines had then sent him to be educated at MIT, after which he continued his government service, working on a multitude of classified programs. They were the kinds of programs that kept the US military far ahead of its nearest global competitor.

But Colonel Gabriel's presence didn't keep her from glaring at the last tattered young man as he wandered in three minutes late and took his seat at the conference table. Jane wasn't used to being kept waiting. By the time she was called in to deal with any situation, it had usually reached the point where it wasn't a matter of whether people would lose their jobs, but how many. She had no problem shutting down an entire facility and had done so on more than one occasion when she felt that the reputation or security of the United States was in jeopardy.

"Thank you all for coming," she said, looking straight at the young man who had just taken his seat. "I'll get right to it." She reached down and fingered a key on her computer, and the projector screen lit up to show a map of Tennessee. A green line appeared on the map, starting at the laboratory and heading into the city of Oak Ridge, and then to the Smoky Mountains. She used a laser pointer to trace the line. "We've verified the path of the particle beam," she said.

She stepped to the screen and put her finger on the line near the laboratory. "The energy field took out one of your visiting Brit engineers, a man named Patrick Ferguson. He had been here since early spring." She pressed a button on her keyboard to advance the slide to show a metal bridge with a hole cored from its center. She paused to gauge the reaction of the men around the table. They looked back at her with blank stares and then turned their gazes to the damaged bridge. She found their lack of emotion irritating and fought the instinct to fire them on the spot.

Jane returned to the map and continued tracing the green line. "Then, it hit a mother and her daughter," she said, now pointing to the center of Oak Ridge. She advanced the slide to show the black Mercedes-Benz. "Six inches of the car were neatly trimmed from the driver's side. The engine was running when they found it." The slide showed the circular slice taken from the car along with a matching indentation that followed into the ground. A rear tire had blown, so the front of the car was tilting towards the sky.

"As you know," Jane continued, "the beam then headed to Clingmans Dome and destroyed that new Q-band tower that everyone was fighting over last year." She showed a picture of the tower. The top half of it was gone, trimmed to a half-moon shape. "If it wasn't an eyesore before, it is now. As far as we can determine, only these three were affected before the beam headed out into space and dissipated. Campers up in the mountains witnessed a loud boom at the same time the accident occurred." She stood there looking at them in silence, hoping for some reaction to help her assess the risk associated with leaving these men in their positions.

"Thank you, Ms. Schaefer," Colonel Gabriel said. "That's all we need for the moment."

"The Propulsion Project is to be shut down until further notice," she replied. She continued looking at their faces with consternation. She had expected to see more remorse on the faces of the men responsible for the deaths of three people.

"You can assure your superiors that we'll not be starting the reactor any time soon," Gabriel said.

"They were clear on this," she said. "The reactor should be dismantled."

"Understood," he replied.

Jane knew enough to catch the dismissive tone in the colonel's voice, but it wasn't her responsibility to question his motives or to gauge the likelihood that he'd follow the orders. She decided to ignore the subtleties and take his response at face value. She disconnected her computer from the projector. "Thank you, sir," she said. She stood and addressed them. "Best of luck, gentlemen." She acknowledged the colonel's nod and then turned and left, shutting the door behind her.

**********

The physicists sat there with their leader, staring and wondering who would be the first to speak. It was Colonel Gabriel. "Well, what do you think?" he asked.

"About what?" replied Brian Palmer. He had looked up in mid thought.

Colonel Gabriel was used to Palmer thinking out loud and knew there was nothing evasive in this young man's nature. He took the question as genuine. "Three counts of manslaughter."

"I'm not saying that we didn't screw up," Palmer said, "but—"

"But what?" challenged the older man.

"These people aren't dead," Palmer said in a matter-of-fact tone.

"Where are they?"

"From what we can surmise," Jacob Cromwell piped up, "when the particles escaped the reactor, the surrounding mass collapsed."

"You think these people disappeared into a wormhole?"

"Our guess is that they were propelled through it as the field collapsed behind them," Cromwell continued. "Like being squeezed through a giant tube of toothpaste."

"Why did your toothpaste head to that tower?" the colonel replied.

"We haven't figured that out yet," David Greer said. "We know it wasn't random, so that's a start."

"Can you get these people back?"

"It's possible," Greer replied. "We have to find them first."

"Where?" Colonel Gabriel asked.

"If Einstein was right," Cromwell said, "the question should be *when?*"

# CHAPTER 4.

## UNCOMFORTABLE FRIENDS

Matt was sitting behind the house contemplating green leaves and trying to press away his headache, when a tall teenage boy interrupted him. "I'm Jeb," he said. "Father asked if you could come help with a horse."

"Sure," Matt replied. He looked again at the green trees. "I'm Matt."

They hurried to the opposite side of the horse barn to join Thomas. He was there with Jonathan, who was the youngest son, another older man Matt hadn't met, and the dog. They faced a horse cornered against a fence. He was a beautiful monster with a rich black coat. "Nice horse!" Matt exclaimed.

"Father calls him Satan...ofttimes," Jeb replied, chuckling. As if he had heard the boy's comment, the horse reared and backed farther into the corner. The dog barked commandingly. The horse settled briefly at the sound, but it didn't take him long to return to his aggressive posture. He snorted, daring them to approach.

Thomas pointed Matt to a spot that would prevent the horse from moving out into the pasture. "Stand yonder," he instructed. "We must get him into the barn. One of his shoes has come loose."

Their efforts lasted for ten minutes as Thomas attempted to bring him under control. The boys and the older man weren't much help; all three looked scared. No matter what Thomas tried, the horse wouldn't let him near, and so he backed away to come up with a new plan. Matt began to grow impatient, so he walked forward. Thomas saw what he was doing and said in a monotone, "Mr. Miller, best to give him space." Despite the warning, Matt continued to walk until he was directly in front of the animal. The stallion gazed down at him with a "Who the hell do you think you are?" expression. Thomas repeated with added urgency, "Mr. Miller, give him space." Matt moved closer, almost as if in a trance, and reached up. As he touched the side of the horse's head, the animal went quiet and stopped rearing.

Matt talked to Shadow like all the animals he saw on the trail. "You're pretty mad...take it easy. We'll get your shoe fixed up, boy." He had practiced this many times on the creatures that crossed his path, trying to convince them to delay long enough for him to observe. Once the horse had quieted, he reached up to his halter and began pulling him towards the barn. There was a sense of relief when he was in his stall and they had shut the gate. Matt walked back out into the pasture with the two boys. "He's okay now. Ready to get that shoe fixed."

The man Matt hadn't met yet now stood in front of him. "Well done!" he exclaimed. "I'm David Taylor, Thomas' brother."

"Matt Miller," Matt replied, shaking his hand.

The boys still looked at Matt in amazement. "How'd you do that?" Jonathan asked. "No one but Father can go near Shadow."

"It felt like the right thing," Matt replied. Thomas opened the pasture gate and they walked back to the house as a group. Matt contemplated the green leaves around him, trying to get his bearings. He couldn't come up with a plausible explanation for why fall had turned to summer.

"Will you be staying for dinner?" Thomas asked.

Matt's head was still pounding, and he was too groggy to think about an alternative. "Sure."

**********

It wasn't as hot in the house as earlier, or he wasn't as sick as he had been. Either way, Matt didn't have the uneasy feeling that he'd collapse as he entered. He could see Mary, David's wife, Faith, and Grace working down in the kitchen. Grace's ice-blue eyes were still vivid in his mind. Thomas pointed him to a chair at the wooden dinner table. The men sat and Mary, Faith, and Grace brought the meal from the kitchen to the table. The boys looked hungry, but no one touched the food, even when the women had taken their seats. The food was passed around dish by dish, but still no one ate. Matt scanned the table for clues as to what he was supposed to do.

Thomas glanced at Matt and said, "Mr. Miller, would you say the prayer?"

"The prayer?"

The boys snickered and he saw Grace roll her eyes.

"To thank the Lord for providing us with this meal," Thomas replied.

"I was born Catholic, but don't go to church," Matt said proudly. "Probably best if someone else handles the praying." There were more snickers, and Matt could see the disapproval in Grace's face.

Thomas took a scolding tone with his family. "We must bring all the sheep home," he said. "Let us bow our heads. Jeb, please lead us." They grew serious, folding their hands, and Jeb began his prayer. Matt followed, looking up occasionally to make sure he was doing the right thing.

*Blessed art Thou, O Lord our God,*
*Who has sanctified us with Thy commandments,*
*And brought forth this food from the earth.*
*Bless us, O Lord, and these Thy gifts,*
*Which we are about to receive from Thy bounty,*
*Through Christ our Lord.*
*Amen.*

There was a roar around the table as Jeb finished and they started to eat and talk. "You should have seen Shadow!" Jonathan exclaimed. "He was awful mad. He threw a shoe and we needed him in the barn. He'd not go until Mr. Miller helped."

This was the first time Grace looked at Matt since they sat. "I thought you had no experience with horses, Mr. Miller."

"You can ride, though?" asked Jeb.

"I don't know how to ride a horse," replied Matt. The boys looked at him like he was from outer space. Matt was beginning to feel like an alien anyway, so their expressions didn't surprise him. "I never had a reason to learn to ride a horse when I was growing up."

"Are you from a big city?" Jeb asked.

"Philadelphia."

"What do your people do in Philadelphia?" asked Thomas.

"I make medicine for a big company," Matt said. "My father drives a cab. My mother lives in New York, though I'm not sure exactly what she does."

"Your family is apart?" asked Mary.

"Yes, my mother left the house when I was twelve," he said. "I grew up with my dad."

"Your mother must be overwhelmed with grief," Mary replied.

"She's fine," Matt said. "Remarried."

They all looked at him in despair.

"No, she's okay," he explained. "We were all probably better off in the end."

This had been Matt's stock answer for questions about his family. The truth was that his parents' divorce had been devastating to him and his sister. His mother had moved out of the house one day when he was in seventh grade so she could go "find herself." He had come home from school in time to see half the family's furniture lifted into a moving van. Since then his mother had popped in and out of his life, mostly on holidays. He had reached the point where he didn't even want to see her at all. He grew weary of her explanation that she had left to "make it better for everyone." He remembered that it hadn't been "better" for either him or his younger sister.

"Are you Amish?" Matt asked.

"Why would you think we're Amish?" Grace said.

"I don't know," he replied. "You don't use electricity, no television, you dress simply." He paused to chuckle. "You all eat dinner together."

Grace looked at him with disdain. "Only Amish act thus in Philadelphia?"

"I think I may not be used to church people," Matt said.

The younger boy, Jonathan, finally changed the subject. "Father, when will you put the new shoe on Shadow?"

"Tonight, while we have him in the barn," he answered. "I desire help." Both boys turned their heads to Matt, obviously not wanting to go near the horse.

"I can," Matt said. "It'll keep me here after dark, though, so I'll need a place to stay another night."

"Mr. Miller, you're welcome to the barn another night even if you don't help," Thomas said.

There was something very confident and caring about this man, which Matt found comforting and intimidating at the same time. For some reason, he felt like he needed to impress Thomas. Matt turned briefly to Grace, expecting some reaction, but she continued with her meal as if not aware of the conversation.

**\*\*\*\*\*\*\*\*\***

After dinner, Thomas led Matt to the barn and Shadow's stall. "He'll be easier to control now that he's closed in," Thomas explained. "Being around the other horses brings out the worst."

"Jeb said you call him Satan," Matt said.

Thomas seemed embarrassed by this. "I guess on bad days he can be a little hard to control. I make sure never to let him alone with Grace or the boys."

"Why do you keep him?"

"You have but to look upon him," said Thomas. "His foals are healthy and strong. They are prized by those who buy our animals." He continued in a joking fashion, "I'd not say this for him to hear, but he's responsible for much of the silver it has taken to build this farm."

Thomas grabbed tools from the bench. "I shall reset the shoe until we can have the smith fix it properly. Keep him calm, and confine him to the corner." Thomas grabbed hold of Shadow's halter and eased him into the back corner of the stall. He motioned Matt to take him by his head. Matt reached up with one hand and began talking to the horse in a soft voice. The horse stared down at him, breathing strongly. He was a warhorse, impossible to transform into a kitten, but he remained mostly still as Thomas worked on the hoof.

The sun was low in the horizon, and it was noticeably darker in the barn by the time they were done. Thomas checked his work and backed away from the stool he had been using. "Hold him until I have put all to rights," he instructed. "He'll want to move after his confinement. We should be near the gate when he is unchained." He picked up the tools and the stool, and then signaled Matt to bring the horse forward. Thomas was already outside the stall when Matt let Shadow go. The stallion immediately came to life, clomping, kicking, and moving from one end of the stall to another as he stretched out his legs. Matt slipped out the gate, and Thomas fastened it behind them.

With his focus on Shadow, Matt hadn't noticed that the rest of the Taylors had appeared and were moving in or out of the barn. One boy entered briefly to fill a sack from a bin and then was off to feed the chickens. Thomas looked around at the activities and, seeming satisfied, motioned to Matt. "The animals must be collected from the south pasture for the night."

Thomas led them out of the barn to another pasture, which was bigger than the one where Matt first saw Shadow. Approximately twenty horses of different sizes filled the pasture, with a few being a good walk away. Thomas opened the gate and pointed. "Can you bring those from the far end?" Matt nodded. He had worked at a relative's dairy farm as a boy and had herded cows, and he didn't think horses would be much different.

"Scout!" Thomas called out. The dog came trotting around the corner of the barn. He looked suspiciously at Matt and growled. "He'll help," Thomas said. "There'll be animals that he pushes to the corners. Start them moving in the proper direction and he'll finish the task." The dog stared up in anticipation. "Scout," Thomas said, pointing to the corral next to the pasture. "In!"

The dog went to work. He ran far ahead of Matt, who walked hurriedly behind. A few horses already walked towards the barn on their own, but others lingered, and the dog barked them into movement or nipped at their legs. Matt found himself in the way of the herded horses and had to jog toward the perimeter of the pasture. Three horses stood at the farthest edge, backed up against the fence. The dog waited impatiently to the side as Matt pulled out the first two, then he stepped in to herd them into the main group. He came back for the third horse and repeated the exercise. Matt watched Scout, impressed by his ability to move the animals towards the barn as an organized group.

A large chestnut horse lingered behind to play with Scout. Matt remembered him as Thunder, having met him in the barn with Grace. At one point, the horse stepped in behind the dog and nearly pushed him off his feet with his head. Scout recovered his footing, circled back around the horse and nipped at his legs. Thunder bounded forward out of reach, then paused to face Scout. They repeated this more than once as they edged toward the corral. Eventually the game ended and Thunder joined the others walking toward the open gate, with the dog following behind in satisfaction.

Thomas, who had been switching around gates and fences, guided the horses into a corral close to the barn. The dog focused on the task until the last horse had entered. Thomas leaned over, patted him on the head, and said, "Good dog." Matt met the dog's gaze, briefly wondering whether he should also pat him. Scout noticed him moving closer and growled, making Matt pull his hand back.

"Cujo," Matt whispered under his breath.

When they'd finished corralling the horses, Thomas checked to make sure there was water in the trough and then he led Matt back towards the house. "Wash up," Mary said as they walked through the door. "There's soap and a tub." She pointed Matt in the direction of the hearth. Matt stepped down into the kitchen, grabbing his pack. There was a basin where he had seen the Taylors cleaning dishes. It had a metal tank above that provided water, along with a drain that led outside. The soap was a crude bar that didn't smell like anything as he foamed it up to wash his hands, arms, and face.

Grace was the last to come in from outside. "Joshua's coat looks the worse," she said to her father.

"We must keep him away from the others until it clears," Thomas replied, "or 'twill spread." The girl nodded in agreement. She briefly met Matt's glance, then looked away and went into the kitchen. Jonathan, the youngest boy, was already there when Matt took his seat at the table. He had a Bible open in his hands and was mouthing the words as he read. He looked up at Matt and returned to his Bible and began mouthing words again. Thomas entered the room wearing a fresh shirt and pants. He looked around and called, "Grace, we're ready to read."

When Grace finally returned, smoothing down her thick blond hair, she was clearly irritated at being rushed. Matt made a point of looking at her as she came in, so as not to seem like he was staring when she sat down. Even scowling, she was strikingly attractive. It was getting dark in the house, so Mary stood to light the oil lamps using a candle she carried in from the kitchen. The lamps gave a brighter light once the globes were set and adjusted.

Thomas spoke as the family bowed their heads. "Let us remember Kathryn, our beloved daughter and sister," he said. "Lord, we know 'tis Your will and we shall never fully understand Your plan. We pray that You care for Kathryn in heaven and help us to heal our sorrow." He went quiet for a moment to regain his composure. "We miss her greatly," he continued in a broken voice. Mary was quietly sobbing. Thomas finished the prayer. "Please bless this family and farm that we may live to accomplish that which You have planned." He paused and said, "Amen."

It was quiet for a time, but then the boys started to get antsy. Mary wiped her eyes and asked for the Bible and began to read from one of the Gospels. It was the story of Jesus walking around collecting disciples. Although Matt wasn't practicing anything as far as religion went, he believed in God and had had enough exposure growing up to consider himself a Christian. In truth, he had grown up jealous of his friends who did have a religion, mostly because it gave them a connection to others in the community that he didn't have. In addition, he had read large parts of the Bible in college, during a time in his life when he was thinking that maybe he should have a religion.

As a scientist, Matt had been encouraged to adopt the Big Bang and evolution as the explanation for everything in the universe. The more he learned about science and the human body, though, the more he'd started to believe that life was not all based on the lucky rolls of random cosmic dice. He suspected that many scientists in his field had a similar perspective, but peer pressure was such that there was nothing to gain by voicing a belief in intelligent design. Even as a child, Matt had always felt that something else was going on in the universe, and he often found himself staring up at the ceiling and asking God for some hint of the plan. His science had never given him an explanation for his instinctual need to look into the sky and wonder if he had some predestined role to play.

The Taylors were religious people and it seemed to give them strength. As far as Matt was concerned, that was something to be respected. He wondered about the story of Kathryn, though he knew this was not the time to ask. He studied them as they took turns reading the Bible. Even the boys were expected to read their sections fluently as their father corrected them impatiently, and so it became apparent to Matt why Jonathan had been practicing. Matt watched transfixed as Grace read her passage of the Gospel. She was a radiant angel in the lamplight. By the time the Bible reached him, Matt was ready to do as good a job as he could. He read with expression and feeling and he noted the look of satisfaction on Thomas's face.

"You've a passion for the word of God, Mr. Miller," Thomas said. "You should find a church when you return to Philadelphia. It's important for young men to worship, even if it seems inconvenient." Matt nodded respectfully, though he couldn't imagine a situation that would make him want to start attending church. Thomas was the last to read. He eventually said "Amen" and the family stood to prepare for bed. Thomas grabbed a lamp and said to Matt, "I'll take you to the barn."

Mary interrupted. "Let me get him a cloth so he can wash for church."

"Church?" Matt said, laughing.

"We attend church on the morrow," Thomas replied. "'Tis the Sabbath."

"Church?" Matt repeated, now more seriously.

"I'll wake you," Thomas added.

Matt was beginning his polite refusal when he noticed Grace watching him out of the corner of his eye. He turned subconsciously to meet her ice-blue eyes, and she looked away. It was enough to make him reconsider. The blue eyes were still in his mind when he said hesitantly, "Sure, it'll be a great story to tell the people back home."

Thomas led Matt to the door and they walked to the barn with a lamp. As hard as he tried, Matt couldn't tell the color of the leaves from the lamplight. It almost seemed now that he had been imagining that they were green. When they arrived at the barn, Thomas lit another lamp with the one in his hand. He set the towels on the wooden bench that Matt had used previously as his bed and handed him the light. "The well is around the side," he said. "You may wash there." Thomas turned and was gone, sliding the door closed behind him.

Matt smoothed the blankets out on the makeshift bed and crawled in. He lay there in the dim light of the low-ered lamp. He was tired, and it didn't take him long to begin dropping off. As he was about to fall asleep, though, he was startled back to consciousness by a loud scratch-ing at the door. Something was trying to get in. The scratching went away for a moment, but ten minutes later it was back. When he heard a growl and a muffled bark, he jumped up out of bed to slide the door open. Scout shimmied past him, trotted over, sniffed the blankets, and hopped up on the foot of what was supposed to be Matt's bed.

"Wait a minute," Matt said to the dog, "that's my bed." Scout put his head down on his paws with his eyes open. It was obvious that he was resolved not to go anywhere. Matt slowly climbed on. There was enough room for him to fit with the dog at his feet. As he stretched out, Scout let out a loud growl, making him pull his legs up. "I guess there's room for both of us," Matt whispered, resigned. He dropped off to sleep, sharing the bed with his new roommate.

# CHAPTER 5.

# MISTER, CAN YOU GIVE ME THE TIME?

Matt was riding on a motorcycle. The wind rushed around him as he accelerated into the setting sun and objects began to blur as he reached the speed of light. His speed increased as he twisted the throttle and the objects turned to ribbons of brilliance. As the ribbons faded, shadows of people and events began to pass around and through him. A few lingered long enough for him to see while others raced by. He recognized the Berlin Wall and saw it crumble, saw Martin Luther King at the foot of the Lincoln Memorial, bombs exploding, and soldiers landing on the beaches of Europe. It began to rain, and the droplets hit his face as he sped forward. They were hot and blasted his face like sandpaper. His face grew warm and wet as the drops fell to the loud beating of a drum. The vision turned to white.

Matt shuddered awake and opened his eyes. Light streamed into them as he turned his head towards a long wet tongue. Scout was licking his face. "Cut it out!" he said, pushing him away. Matt heard knocking on the barn door and lifted his head from the bed.

"Mr. Miller, Father said you should wake," Jonathan said from outside. "We leave for church within an hour."

Matt called out, "Thanks!" He heard the boy shuffle away.

Matt sat up and the searing pain returned to his head. He had hoped that a night's rest might soften his headache, but it was as bad today as yesterday. He sat there immobilized for a moment, waiting for it to subside. Once it dimmed to a mild throb, he pulled his pants and shoes on, went to the well for water and brought it back to the barn. The cold water made him shiver as he brushed his teeth, washed and shaved. He dried thoroughly with one of the coarse towels Mary provided and then put on the last clean shirt from his backpack. Matt zipped the pack and picked it up to place safely in the house.

He closed the door of the barn behind him, then thought better of it and opened it slightly for the dog. He walked to the house and knocked. He turned while on the porch and was reminded of the green leaves, but before he could think about it, the youngest boy, Jonathan, answered. Both Mary and Thomas stood in the entryway dressed for church. Matt knew as soon as he saw them that his clothes wouldn't be appropriate. Mary was in a long pale dress with a bonnet and Thomas looked like one of George Washington's colonial soldiers. All he needed was a wig and a triangle hat. Matt had to keep from laughing at his black shoes and stockings, wondering again what kind of religion required them to live such antiquated lives, but he kept his thoughts to himself.

"I'm ready for church," Matt said tentatively.

Mary and Thomas inspected him with some scrutiny, but didn't say anything. Grace, who had quietly walked into the room, had no problem vocalizing their opinions. "He can't go thus!" she exclaimed. Grace too was wearing a colonial dress with white ribbons that matched those in her hair. Matt imagined her as one of the Southern belles of old.

Mary spoke to Thomas. "Will's clothes may fit him." Thomas nodded and Mary motioned for Matt to follow her up the steps. At the end of a hallway with a number of bedroom doors, Mary opened the last door onto a room containing a narrow bed and a wooden wardrobe. A cloudy glass window flooded the room with light. Mary stepped to the wardrobe, opened the doors, thought for a moment, and pulled out a coarse white shirt, a blue vest, a blue jacket, and a pair of short grey pants, and laid them on the bed. She seemed deep in thought and didn't move for a long time, then went back to the wardrobe and pulled out long socks. She reached under the bed for black shoes. They were dusty, so she found a cloth to wipe them. "These may fit," she said. "They were always too large for Will."

"Who's Will?" Matt asked.

"My son. He lives in town." She had a hint of pride in her voice. "Try the clothes. You're welcome to some of the others if these aren't appropriate." She lingered for a moment and then left, closing the door behind her.

Matt scrutinized the outfit. This would be something to tell his friends when he went home. He pulled the socks on, then the long woolen shorts, which went past his knees and covered up the ends of his socks. He buttoned the bottoms of the shorts so they would stay in place. He put the shirt on and tucked it in around his waist, then walked to the bed and sat down to examine the shoes. They were black with a silver buckle, like something the pilgrims wore during the first Thanksgiving. His feet slipped easily into them, and he reached down and fastened the buckle. After this came the vest, over which he buttoned the jacket. He ran his hands over the clothing to smooth it and then placed his own clothes in a bundle near his pack. He sniffed the air, noticing that he now smelled like the oiled wood from the wardrobe.

Matt pulled his phone from his pack and looked for service, but there were still no bars. He propped it up on the wardrobe and began to shoot a video to record his latest adventure. "I don't have phone service right now," he said to the phone, "but as soon as I do you guys are going to die laughing. I got mixed up with some religious people. They dress in old-time clothes for church. I'll take some pictures when I'm there." Then he added, "And you always wonder what I do on my hiking trips." Matt was sure he finally had something to beat his "being chased up a tree by a bear" story. He slipped the phone into the pocket of this jacket.

As he was walking out of the room, he thought twice, went to his pack and unzipped the side pocket of his shaving kit where he kept his watch. He was surprised to see that his ring was there from his last trip home to see his father. It was a fourteen-karat-gold monstrosity with a ruby-red stone and four diamonds. He put them both on and stood there for a moment, watching them sparkle in the sun that seeped through the windows before walking downstairs to join the Taylors.

"The clothes fit," exclaimed Mary, proud of her selections.

"Now you're dressed to worship the Lord," Thomas pronounced. He turned and led them all out the door. There was already a wagon in front of the house hooked up to two black mares. "These are two of Shadow's off-spring," Thomas said. "We'll show them in town after church." Thomas lingered as he rubbed one of the horses on the neck. They were identical, muscular and beautiful, with rich red-black coats that shimmered in the morning light. The horses had perfectly proportioned white socks on all four legs that contrasted dramatically with their coats.

Jonathan hopped into the wagon first and Matt sat next to him on the bench. Grace and Mary took seats immediately in front of them and proceeded to adjust their dresses so they didn't pull their shoulders down as they sat. Thomas and Jeb got up into the front seat. Thomas slapped the reins and the wagon jerked forward.

They drove about twenty minutes along a dirt road. The green leaves on the trees were now too obvious for Matt to ignore, so he spoke. "The leaves are green."

"We've had a lot of rain," Thomas replied. "Not like last year when they were brittle from drought."

"I mean for this time of year," Matt said. "They should be falling."

"Are they falling already in Philadelphia?" Jonathan asked.

"Of course they are," Matt said. The boy gave him a strange stare.

They approached a town filled with wooden buildings. Matt could see a white church steeple. A bell rang in the distance with a regular rhythm, getting louder as they approached. From the outskirts, the town resembled one of those Wild West towns in the Midwest that they build for tourists. There was a whitewashed tower with a name that Matt couldn't read, so he tapped Jonathan's shoulder and pointed. "What town is this?"

"Richmond," Jonathan replied.

"Like Richmond, Virginia?"

"Yes, Richmond."

They pulled up to the church through the split rail fence into a clearing in the yard. Other teams of horses came in through the eastern gate of the churchyard. The wagons were various shapes and sizes, some being led by up to four animals. Thomas stopped their vehicle as they came around to the front of the church and both he and Jeb hopped out to help the women down. There was another round of dress smoothing and setting everything in place. Matt saw Jonathan step over the side of the wagon and he followed him to the ground.

Thomas asked Jeb, "Can you handle them yourself?" Jeb nodded and jumped up onto the wagon, shook the reins, and proceeded to move it to where the others parked at the side of the church. This makeshift parking lot, relatively empty when they arrived, was rapidly filling with many types of wagons and buggies.

Jeb finally appeared from a group of people who had gathered around the vehicles. These people all seemed to be staying with the horses. They were black and Matt immediately felt that something was odd about this. He looked around at the people entering the church. There were many types, most elaborately dressed, but they had one thing in common: all were white. He looked over at the wagons and tapped Jonathan on the shoulder. "What about those people?" He pointed. "When do they go into church?"

"They have a separate church," the boy replied matter-of-factly.

"Are they a different religion?" Matt asked.

"They're Christian," the boy said.

Matt pressed. "Why don't they go to this church?"

The boy frowned. "Slaves are kept outside."

Matt knew he had misheard. "Say that again."

"The church people have rules," the boy said, louder this time.

"Did you say slaves?"

"You can't just let slaves enter the church!" the boy exclaimed.

Matt looked around and it all came together for him. The green leaves, the people, wagons, buildings, even the Taylor family—all were pieces of the same puzzle.

"Another question," Matt said, putting his hands on the boy's shoulders.

"Sure, Mr. Miller."

"What year is it?"

"It's seventeen hundred and sixty-two." The boy said it in a singsong fashion, as if he had practiced it before.

"Did you say seventeen sixty-two?"

"Seventeen *hundred* and sixty-two," the boy corrected.

# CHAPTER 6.

# TOOTHPASTE, PART III

"So what's the latest?" Colonel Gabriel asked. The physicists were having their first meeting since receiving instructions from Jane Schaefer to disassemble the Chernenko-Einstein reactor.

"We found a connection between the people and the tower," Brian Palmer said. "The particle stream formed between the signals from their quantum phones as it made its way to the cell tower."

"How'd you figure this out?" Gabriel asked.

"They went through my phone first," Palmer replied. "It seemed like too much of a coincidence."

Jacob Cromwell smoothed the red hair from his face and spoke. "We believe there was another person," he said.

"A fourth?" exclaimed Gabriel.

"We did a high-resolution analysis of the particle path," Cromwell explained. "It changed slightly between here and the tower. There was another phone near Clingmans. Probably a hiker."

"A hiker?"

"A missing person report was filed for someone named Matthew Miller," Cromwell replied. "He was hiking the Appalachian Trail."

"Let's focus on him for now," Gabriel said. "How do we get him back?"

"We're working on that," Cromwell replied. "We think we figured out a way to contact him."

"Send him a note through a wormhole?" Gabriel said.

"Better," Cromwell replied. "We can send him a text."

# CHAPTER 7.

# BIDING TIME

---

Matt's thoughts raced. He looked around for an escape route, but he had already made the mistake of moving with the Taylors into their church box, where they were followed by what had to be the most portly family in Richmond. Unless Matt wanted to physically climb over someone, there was no way he could get out.

He made a conscious effort to calm himself and think of a plan, but in the turmoil of his thoughts, none came. He briefly entertained the idea that he was dreaming, but too many things had happened since he woke up in the barn for him to believe he was unconscious. The people and the smells were real. He reached out to the wooden barrier in front of him to feel its heavy bulk. This was 1762! He needed to find out how he got here and how he was going to get back.

Matt grew increasingly impatient. Grace glared at him, annoyed with his fidgeting. By the time church ended, he had been through about one hundred possible scenarios as to what to do next, and nothing was even close to sounding right. He followed the Taylors outside to the gathering in the churchyard, purposely lagging behind and hoping to make his quick escape, and then took the first opportunity to step to the side. Once he was free of the Taylors, though, he realized that there was no place to go. He'd have to ride back with them to retrieve his pack. The only plan that seemed plausible was to go to the place they found him and hope there was some clue for returning to his own time.

Helpless until he got his pack, he resigned himself to standing there and watching people greeting one another like characters in some bizarre colonial charade. From this vantage, he saw Mary and Grace talking with other families, and Thomas across the yard with a younger man. A few well-dressed families with tall young men stopped to make conversation with the women. Each one lingered long as he greeted Grace. Matt tried to gauge her reaction as they took her hand, looking for a hint of a smile or a frown, but he saw nothing that gave her away.

After the third man introduced himself, Grace looked up and caught Matt staring. He immediately looked away. He tried to correct this by looking at her again, but she had gone back to her greetings.

"Fine clothes, sir," someone said to him. "Very fine clothes." Matt turned from Grace to look into the eyes of a man who was now standing in front of him. Thomas had been talking to the same man.

"Uh," Matt said. "What?"

"If you can pry your eyes from my sister, I should like to introduce myself. I'm William Taylor."

"I'm Matt," Matt stammered. "Matt Miller." He reached out to shake William's hand. They were about the same height and build. "William Taylor? You're the missing son."

"Not quite missing," he said. "Just away in the big city."

"You live in Richmond," Matt said. It was an innocent confirmation on Matt's part. In his time, Richmond was a big city. He had no idea what the population was in 1762.

"Most nights," Will answered. "My father says you've taken Scout's barn."

"We've worked it out," Matt replied.

"Did they find you drunk under a bridge?"

"I wasn't drunk. I was sick."

Will nodded. He didn't seem like he cared one way or another, almost like he thought it was a funny story or maybe a badge of accomplishment. "Too bad," he replied. "A drunken scoundrel might be the perfect diversion for my family." Matt looked at him questioningly, but Will didn't elaborate. "So, why are you sleeping in our barn and wearing my clothes?"

"The barn's where I woke up, and your family didn't think my traveling clothes were appropriate for church." He made an effort to sound as rational and calm as possible. "I have no idea how I wound up under that bridge, but I'm not a drunkard."

Will grinned. "I believe you. Where are you from?"

"Philadelphia."

"Ever been to Richmond?"

"This church is the closest I've come," Matt replied. "Aren't we in Richmond now?"

"The city proper is down the road."

"Do you know this Bonner Bridge where they found me?"

"Sure."

"Can you take me there? I lost something."

Will nodded. "I have errands, but it's on the way."

Matt glanced around, looking for the Taylors. He was worried about retrieving his pack and getting on the road before dark. "I'd like to make it back to your farm with plenty of daylight," he said.

"I'll be going for supper," Will replied.

Matt thought about it. On one hand, he wanted to go back as quickly as possible and get his stuff, but on the other, he might not have another opportunity to look for clues. "Okay, let's go," Matt finally said.

"You'll be fine not to gaze upon my sister until supper?" Will smacked him on the shoulder and laughed.

Matt rolled his eyes. "I'm ready when you are."

"I'll tell Father." Will wandered off into the crowd to find Thomas. Matt's head was starting to hurt, so he put his fingers up to his temples and pressed, trying to massage the pain away.

"You've met my brother." Matt turned to see that Grace had stepped in behind him. She stood there, looking at him with those piercing blue eyes. "He could probably take you for a drink in town should you desire." She was trying to be funny, but something about her smirk was infuriating.

Matt looked back at her plainly. Who did she think she was? In as calm and even a tone as he could muster, he said, "I'm no drunkard." Matt looked away into the crowd, anxious to be away from this woman. He had many more things to worry about now than dealing with some spoiled country princess. He stepped away from her to look for Will, but she reached out and grabbed his arm. She caught herself, dropped her hand, and looked around to see if anyone had noticed her indiscretion. Satisfied, she stepped close to Matt so she could speak in a lower voice.

"I'm sorry." She seemed sincere but amused. "I was wrong to judge." The apology was forced and it reminded Matt of when he got in a fight in grade school and was asked to apologize to a boy who moments before had been intent on beating him silly.

"I'm taking him away from you, sister!" Will said. He had walked up to them unseen. "We're going into Richmond for the day. We'll return this afternoon."

"Don't be late. You know how Mother worries," Grace scolded.

"Punctual as ever, good sister!"

Grace nodded knowingly, said goodbye to both of them, and walked away, not looking back.

"Desired one more glance?" Will said to Matt, smiling.

"Let's go to Richmond," Matt replied.

# CHAPTER 8.

# RICHMOND

---

Will gave the slave tending to his horse and carriage a coin and pulled the carriage forward. He motioned for Matt to follow and then, reins in hand, hopped up. Eventually Matt discovered a footstep and hoisted himself onto the seat next to Will and they headed out of the courtyard.

"Ofttimes it seems that church will go on forever," Will said. "Richmond will be better, with all the beautiful ladies out in their finery." He paused for a moment and then quipped, "If you can keep my sister from your thoughts." He broke out laughing and Matt couldn't help but smile. It was like having a fraternity brother sitting next to him.

They didn't ride very long before coming to a bridge. Pointing, Matt said, "Is this Bonner Bridge?"

"Desire a nap?" Will replied.

"Funny. Do you know exactly where they found me?"

Will pointed. The bridge was an archway over a wide creek, built with mortared grey stones. It was surrounded by dense brush and weeds, but had a path that led down to the water. Matt jumped off the wagon and said, "I'll be back." He walked quickly down the path and under the bridge. When he reached the creek, he stood there looking for some clue of his arrival. He had expected the area to look charred, like a rocket had launched there, but the area didn't seem disturbed except for some discarded fishing equipment. He kicked at a pile of rusty hooks, string, and an old cork bobber.

Matt walked around the clearing, probing bushes and stomping on the ground, but there was no evidence of a magic portal that would take him home. "This can't be it," he said aloud. Then a splash of red in the bushes caught his eye. He reached down and found the red bandana that he had worn on the trail. He examined it briefly to appreciate what it represented and then stuffed it in his pocket. He looked up into the sky, shaking his head angrily. "Now what am I supposed to do?" He walked slowly back to the wagon.

"Did you find what you lost?" Will asked.

"Not really," Matt replied, not knowing how else to answer. "Maybe I'll find it in Richmond." Will gave him a puzzled look, waiting for some explanation, but Matt was already lost in thought, trying to plan his next move.

*********

The city surrounded them once they were in view of the James River. Houses lined the bank of the waterway and barges floated gently on its surface. Will drove the carriage across a bridge and they entered a business district. Some shops were painted white or green and others had the grey of natural wood weathered over a number of years. All were closed for the Sabbath. They passed by an apothecary and then a silversmith shop with a sign that said "Jewelry for Sale" in the window. Will stopped the cart in front of a shop with a sign that read "Samuel L. Smith, Accounting." The sign hung on hinges, flapping in the wind. Will jumped down and looked up at Matt. "I'm to meet with Mr. Smith," he said. "Do you mind waiting?"

"Not at all," Matt replied. "Is it okay to leave the carriage and walk around?" He welcomed the opportunity to window shop.

"'Tis a reputable street," Will replied.

Matt glanced at the silversmith shop and then down at the college ring weighing heavy on his finger. It had been a graduation present. His father had driven a limousine for a man who owned a jewelry conglomerate. It was custom-made, an elaborate carving of Independence Hall on one side and an eagle with a snake in its talons surrounded by decorations on the other. In its center was a ruby-red stone. Matt was mostly sure it was not a ruby but had never asked. Surrounding the stone were the words *Philadelphia, Pennsylvania*, and these words each had two diamonds interspaced between the letters. His dad drove this jeweler for a long time, so Matt was sure that he bought it at a discount, but he suspected it was still worth a sizable sum. Matt didn't generally wear the ring unless he was visiting his father. It had been in his shaving kit from his last trip. Matt was sure his father would understand the need to sell it, considering his present circumstances.

Will was turning to leave when Matt said, "Do they take dollars in this town?"

"What are dollars?"

"That's what I thought," Matt replied. He held the ring up for Will to see. "How much do you think this ring's worth?"

Will stepped closer, then stared into Matt's face as if he didn't want to answer. "You sure you want to sell?"

"It's a large ring that I don't normally wear," Matt replied. "I need money to travel home."

Will motioned for Matt to hand him the ring, then hefted it a couple of times. "It's heavy," he proclaimed. "Those diamonds are worth something. What's this structure on the side?"

"It's called Independence Hall."

"Independence Hall?" Will looked at him. "Independence from what?"

"That's a long story for another day."

"Maybe forty pounds, give or take a few shillings."

"How many shillings in a pound?"

Will stared. "Are you serious?"

"I don't deal with money much," Matt replied. "Another long story."

"Twenty shillings to a pound."

"You don't have a shilling and pound on you?"

"You're a colonist and you've never seen money?" Will asked in disbelief.

"We barter where I come from," said Matt. He was pleased with himself for thinking of this very plausible explanation on the spot.

"Seems exceedingly strange," Will replied. He slowly pulled out a gold and a silver coin and gave them to Matt. Matt turned them in his hand, feeling their weight. He returned the coins and watched Will secure his purse.

"One other thing," Matt said. "How much would your father charge for one of those prize horses he had on the wagon today?"

"Mares sired by Shadow," Will replied. "Anywhere from forty to sixty pounds each."

"Thanks," Matt said. He hopped off the buggy, walked to the silversmith and knocked. A middle-aged man with greying hair and some girth around his belly immediately came to the door. He wore a weathered leather apron that covered most of the front of his body.

"'Tis the Sabbath," the man said through the door window. "I'm closed."

"Can you open the door?" Matt motioned towards the knob.

The man opened it a crack. "Son, I told you, the shop is closed."

"I have a gold ring with diamonds you should look at," Matt said, holding the ring up.

"That a ruby?"

"I think," Matt said, thinking it best to play dumb.

The silversmith opened the door wide. He rubbed his hand on his pants under the apron and stuck it out. "Jacob Berkley."

"Matt Miller, from Philadelphia," Matt said as he shook his hand. "I'm a friend of William Taylor, who is doing business across the street."

"Doing business?" Berkley said. "More like indentured servitude." Matt smiled and shrugged. "Come and sit," the silversmith said, pointing to a stool at a counter near the window. "Let's see it in the light." Matt sat down opposite the silversmith, facing the street. He handed the man his ring and watched as he went through the examination. "Those diamonds must have cost a pretty penny," Berkley said. "Where did you get this?"

"My father gave it to me as a graduation present."

"Graduation?"

"The Philadelphia College of Science."

"Makes sense," Berkley said, looking at the words around the red stone. "Is this the Pennsylvania State House?"

"Yes," Matt answered. Most people from Philadelphia knew that Independence Hall was originally called the Pennsylvania State House.

"The carving is intricate." Berkley continued turning the ring in his hands and then said, "I'll give you thirty pounds."

"Thirty pounds!" Matt exclaimed with contrived indignation. "I could never take that!"

"Thirty-five, but 'tis the highest I will offer."

"Surely you can do better than that."

The silversmith looked back at him and was silent.

Matt said, "It's a heavy gold ring with four diamonds and a ruby. Surely it's worth more than thirty-five pounds."

The silversmith inspected the ring again. "I can't be sure it's a ruby. The stone has a flaw."

"What flaw?" Matt asked.

"'Tis dull in its center," the silversmith said. Matt motioned and the silversmith handed it back. He looked carefully at the red stone.

"Oh," Matt said, remembering. "The crown."

"The crown?"

The Crown Jewelry Company did a lucrative business providing championship rings for the world's wealthiest sports teams, but two challenges had appeared the last couple of years. Cheaply made rings from legitimate overseas competitors were appearing on the market as well as replicas of real championship rings from unknown manufacturers. The company had tried to combat this by using a laser to engrave a crown on the underside of the primary stone. It was etched in a way that made the crown float in the very center of the gem. No other manufacturer had figured out how to replicate this proprietary process. It made for a dramatic effect once you knew where to look. Matt handed the ring back to the silversmith.

The silversmith put a magnifying glass up against the ring to look at the stone. "That's incredible," he said. "'Tis a likeness of the king's crown. How'd they do this?"

"Looks like it took a lot of skill to get it right," Matt replied.

"You tell me a price," the silversmith said.

"Sixty pounds," Matt replied.

"Sixty pounds?" Now it was the silversmith's turn to be incensed. "Too much."

The silversmith returned the ring and Matt stood up to leave. He hadn't even thought of the whole "king's crown" angle, and maybe the stone was a ruby. It could be worth much more to the right customer.

"Sit, please." The silversmith motioned to the chair. "I don't know how much your ring is worth, but I'm not willing to front sixty pounds. I have a number of contacts in the Pennsylvania government. These men are wealthy collectors of luxuries and adamant supporters of the king. They may be interested in such an item. Let's arrange a consignment. If it sells, we split the proceeds by half."

"Half?"

"Half," the silversmith repeated. "I have associates who deal in silver pieces and jewels. You would never convince a stranger to purchase this ring."

Matt saw the truth in this, but he needed money now. "I think half is a lot. I want some small sum up front, with the rest split fifty percent once the ring is sold. How about ten pounds in advance?"

"Seven is as high as I'll go."

"Fine," Matt replied reluctantly.

Berkley went to the back room and returned with parchment and a quill pen. As he sat, he glanced out the window and noticed that Will was now sitting in the buggy. The silversmith went to the door and motioned.

"Good afternoon, fellows," Will said, entering the shop.

"Mr. Miller wants me to sell his ring," Berkley explained. "We desire you to be a witness to the contract." The silversmith sat down and wrote out the terms, then copied them exactly on a separate piece of parchment. Both men signed each document and Will signed as a witness. When they finished, Berkley went into the back room and returned with seven gold coins for Matt. Matt fingered the coins to determine their authenticity and then laughed, realizing he had seen them for the first time only an hour before. He accepted the coins, folded up the agreement, and placed it all into his inner jacket pocket.

"Where do you take lodgings?" the silversmith asked.

"I'm staying out at the Taylor farm," Matt replied.

"How shall I contact you?" Berkley asked.

"I'll let Will know where I'm staying when I leave," explained Matt. Matt turned to Will. "Would you mind? I'll buy you dinner in town once the ring sells."

Will nodded and said, "I look forward to a costly meal!"

"Keep your distance from this one," Berkley said, smiling, "unless you care naught for that gold in your pocket."

The three men shook hands and Will and Matt left the shop. Matt was seven pounds heavier.

# CHAPTER 9.

# LUNCH AT KING'S TAVERN

---

They drove slowly past shops and homes, stopping often to let people, horses, and buggies pass. Will was in no hurry and took every opportunity to shout and wave heartily at people gathered outside. As promised, a number of these revelers were beautiful young women in formal attire. Will stopped the carriage in front of a crowded churchyard. "Let's say good day," he said. Will tied the reins, slipped to the ground, and motioned for Matt to follow.

They walked into the courtyard and people turned immediately to greet them with boisterous handshakes and smiles. "Meet my new fellow, Matthew Miller," Will said. "He's visiting all the way from Philadelphia." It was initially awkward for Matt since he had no idea whether he was supposed to kiss the women's hands. He watched Will and realized that in most cases he was only lifting hands as he talked, rather than kissing. One exceptionally beautiful woman named Graine Martin lingered with Matt after finding out where he was from. She complimented him on his accent and asked many questions about Philadelphia and how he made his living. She seemed to grow increasingly interested as they spoke and gave the distinct impression that she was flirting.

Will wandered over after a while to join their conversation. "Good afternoon, Miss Martin," he said. "I see you've met Mr. Miller."

"He's been gracious enough to tell me about his life in Philadelphia," she replied.

"Hopefully not bragging about his wealth," Will said, winking at Matt.

"He's been gracious and very humble."

"We are on our way to a meal at King's Tavern," Will said.

Graine turned back to Matt. "Mr. Miller, I've enjoyed speaking with you." She offered her hand higher this time, so Matt leaned down and lightly kissed it. He waited with some humor for her to slap him, but only a smile came. She curtsied and said, "Until we meet again." She turned with a swish of her dress and walked back into the crowd.

"I think she likes you," Will said as he motioned for Matt to follow him back to the buggy. "A beautiful lady, but I pity the man she marries. She spends her father's money like there's no end."

*********

It was a ten-minute buggy ride to King's Tavern. There was a picture of a crown on the tavern sign, which Matt thought an appropriate omen. The quiet façade didn't prepare him for the merrymaking that hit them as they entered. The revelry was entirely unlike the sedate impression he had expected of Colonial America on the Sabbath. People crowded along a long bar, and in the corner, a flute player, a drummer, a fiddler, and a singer with a stein in his hand played a bawdy song about French girls.

Will waved to the bar as they entered and a woman rushed over to greet them. To Matt's surprise, she headed straight to Will and planted a kiss on his cheek. She took him by the hand and guided them deeper into the tavern. No one paid any attention, so Matt assumed this was a common occurrence. They moved past the band and into the back room. The noise was less intrusive there and Matt could hear conversations around them. The woman pointed them to their table and they sat down.

"Is there a menu?" asked Matt.

"A menu?" questioned Will.

"A piece of paper that lists the food and the prices."

Will looked at him questioningly. "There's no, what did you call it, *minu*? Is that French? You can usually rely on the cook here to make an excellent meal of what they've selected this day." He smiled in anticipation. "You want ale?"

"It would make me groggy," Matt said. "I don't want to tour Richmond hazy and tired."

"I thought you were a drunkard." Will had a smile on his face.

"As does your sister," Matt said, resigned. "But it isn't true." *Not entirely, anyway.*

The server returned and Matt noted her traditional English barmaid's outfit. He could attest now when he returned to his own time that the buxom bar wench wasn't just a Halloween costume.

"Two ordinaries?" she said.

"Perfect," replied Will, "and also, a flagon of fresh water and a pot of tea." The waitress quickly brought the water and tea. Matt wasn't normally a tea drinker, but from the first sip, it was like a magic elixir. He hadn't had caffeine in a few days and the withdrawal was adding to the pain in his already throbbing head. When they finally were able to speak again, Will said to Matt, "How do you fancy our farm?"

"I like it fine," Matt said. "You have a good family." Matt remembered Will's previous comment about his family at the church, so asked, "Why do they need a diversion?"

"From their grief, of course," Will said. "Father has been overwhelmed since my sister Kathryn died. He'd never allow it, but he favored her. He's not accepted her passing and he's not used anyone the same since."

"How old was Kathryn?" Matt asked.

"Twenty-two and pledged in marriage."

"It must take a while to recover from the death of a child."

"They reenact it every evening. I loved my sister Kathryn as anyone." He paused. "Ah! What is it that you say? It's a long story."

Matt changed the subject with a question. "Why did you move to Richmond?"

"I'm apprenticed to an accountant," Will said.

"You decided not to go into the horse business?"

"Nothing of the sort," Will replied. "When I finish, I'll return and continue what Father has begun."

"Will!" a pretty redhead exclaimed from across the room. She approached the table and to Matt's surprise immediately sat down on his lap and made herself comfortable. Her hair tickled Matt's face and it was hard not to breathe her in. She smelled good.

She looked across the table at Will and said, "Who's your fellow?"

"His name's Matthew Miller," Will said. "He has lodgings at our farm."

"Is he wealthy?" she asked.

"Yes," Will said to the girl, "but he may be pledged to my sister." He chuckled.

She looked over her shoulder at Matt and said, "Yet another obsessed with Miss Taylor. I'm sure there's something that can be done."

"You're welcome to try, Ciara." Will laughed.

"Does he talk?"

"I talk," Matt said. "I've been too overwhelmed by your beauty."

"What accent is this? Are you very wealthy, Mr. Miller?" She said it in a seductive tone, but before Matt could think of a sufficiently witty reply she was off his lap and walking away. She looked over her shoulder and purred, "If you gentlemen should want for anything, please." Then, she disappeared from the room.

"First generation Irish," Will said. "That's truly one where you must watch your purse." Matt reached down without thinking and fingered the coins in his pocket. The commotion had attracted some attention, and Matt saw a man who was eating with two younger men stand up and head toward their table.

"How does it, Will?" he said, extending his arm. Will stood and shook his hand.

"Quite well, Nathan. This is Matthew Miller from Philadelphia." Matt stood to shake.

"Nathan Payne," he said before Will could introduce him. "What brings you to our fine city, Mr. Miller?"

Matt couldn't quite put his finger on it, but something about the man's manner told him that there was more to his question than good will. His instinct told him to say no more than necessary. "I'm visiting the Taylors while I do business here in town."

Nathan looked at him and Matt thought he'd say something else. He seemed to reconsider and said simply, "Welcome to Richmond. I hope your traffic brings much prosperity to our city." He returned his attention to Will. "How does your father?"

"Father is well, as are the others. Kathryn fills our thoughts."

Nathan went silent then. "Ours too," he replied at last, then was quiet again, and Matt wondered if he had gotten this man wrong.

"How does the trade?" Will asked.

"Better than ever. We've forty new animals and three additional Negros."

"Excellent," Will replied. "We have many new foals."

"When will your father sell me that black monster?"

"Bring your mares over anytime," Will replied.

"Oh! For the price your father asks, I should have the whole horse."

"Your welfare weighs heavy upon Father," Will joked. "The malice of that animal is beyond your means."

"Trust me, lad, we can command him," Nathan said. "Thomas has my offer. He could expand his stock by half with the price I should pay."

"You must convince Father," Will replied.

"Ah, 'tis true," Nathan said with resignation in his voice. "Give him my regards." He turned to Matt. "Mr. Miller, 'twas good to make your acquaintance." He shook both their hands and headed back to his table as the barmaid brought their meals. Each plate had a tiny chicken, roasted potatoes and carrots.

"Who was that man?" Matt asked finally.

"He and my father were in the horse trade long ago," Will explained. "They went their separate ways."

"He seems very interested in Shadow."

"You know Shadow?"

"I helped get him into the barn. He's a beautiful horse."

"He has his use," Will said, shrugging.

"How many people are interested in Shadow?"

"Most farms inquire," Will replied. "Anyone who sees the foals recognizes that his is a champion bloodline. He's one of the few means by which we can prosper."

"Why wouldn't you be able to prosper without him? Your father has plenty of horses."

"We can only attract buyers willing to pay a premium," Will explained. "Our labor costs are heavy."

"Labor costs?" Matt questioned. "I've only seen family working your farm."

"Father refuses slaves, so we can't grow as we should," Will said. "Next week, the hay must come in and we'll hire townsmen."

"What's the problem?"

"Many will be new and need instruction for the simplest of tasks. Some may be looking for trouble. Either way, it's no way to run a business."

"So you don't agree with your father?" Matt asked. He focused on asking the question with no judgment in his voice. He knew they were a century away from a human rights struggle that would result in the death of over six hundred thousand Americans.

"I know not," Will replied, "but this is certain: we rely overly on Shadow and a few other animals. Horses like Shadow come along once in a lifetime and 'tis evident he shall soon kill himself or one of us. We must plan for the day when that horse is dead."

"Your father can handle him. I'd not write him off so quickly." Matt could remember the look in Shadow's eyes. He tried to be convincing, but he wasn't sure.

Ciara interrupted. "How does it, gentlemen?" she asked, looking down at their empty plates. "Something stronger in your cups?"

"I don't believe," Will said. "We are to the farm this afternoon."

"What about you, Mr. Miller, can I get you any-thing…anything at all?" She purred when she talked.

"Ciara, go away," Will said. "You don't want to give Mr. Miller the wrong impression, do you?"

She reached out and touched Matt's chin. "No, never."

"Ciara, how much? Mr. Miller and I have business," Will said impatiently.

"Three shillings," she replied. Matt pulled out a coin and handed it to her. She left briefly and returned with a handful of change. She gave a friendly smile to both of them and said, "Thank you, gentlemen, come back soon," and was gone. Matt reached over and left two shillings on the table.

Will looked down at the coins and then at Matt. "We must make our escape before she charms you, marries you, and empties your pockets," Will said, laughing.

As they got up, an older grey-haired man entered the room. He noticed Will and walked up, offering his hand. He faced him directly and didn't pay any attention to Matt. "Has your father considered my offer for the twins?" he asked.

"One hundred fifteen," Will replied, smiling.

The man noticed Matt and reached out his hand. "William Hancock."

Matt shook his hand. "Matt Miller. I'm staying at the Taylor farm."

"What do you think of these horses, Mr. Miller?" The question took Matt by surprise, so he took a moment to think. The man didn't wait for his answer before asking another question. "Would a Christian family spend this exorbitant sum on such a luxury?"

Matt answered sincerely. "I rode with these horses this morning. They were glowing," he said. "They seem to be black sometimes, but when they capture the sunlight they shimmer deep red; it's amazing. Besides, they are from strong and healthy stock. No man would regret owning these horses."

Hancock was quiet for a while, thinking. "One hundred five is my offer. 'Tis a king's ransom." He added, "Your father is impossible to deal with these days."

"I know," Will proclaimed. "I don't think he even wants to sell. He put them on the wagon for church today. They rode into town to visit John McKinley."

"John's looking to buy horses," Hancock said. "He's not trying to sell them to John?"

"Only a friendly visit," Will replied.

Hancock threw his hands in the air. "Fine. Tell your father I'll pay the hundred and fifteen!" he exclaimed. "I'll be out to pick them up tomorrow."

"They're yours if they're not already sold," Will said. They shook hands and the man walked back to his family. He smiled the whole way back to his seat, seemingly very happy with his purchase. As Matt watched their negotiation, some subconscious signal caused him to look over at the table where Nathan Payne was sitting with the two other men. He saw Nathan was staring at Will as he finished the deal with Hancock. Nathan looked directly at

Matt. Their eyes met for longer than was comfortable and then Nathan turned away with a pensive look on his face.

"I guess we'd better hurry to the McKinleys'," Will said. "Father won't get one hundred fifteen from John. This much is certain."

Matt shrugged and said, "What else do I have to do?"

"Let's say good day to Nathan and his boys," Will said. "Competition or no, no one should believe there's a problem between us. It's bad for business." Will led Matt to Nathan's table and reached his hand out. Nathan stood to shake. Matt followed, nodding goodbye. The younger son, Paul, stood up and shook his hand, saying "Good day." The older son, Levi, took his time. He slowly extended his arm to shake Will's hand.

"Your hands are soft," Levi proclaimed. "You should return to farming."

"You're working hard enough for the both of us," Will replied evenly and coldly. "But if you're idle these next weeks, we could use help with the hay."

Matt saw Levi's face redden. "I'll leave that to whatever clowns you suffer this season," Levi said sourly.

Will ignored him. "This is Matthew Miller. He's come from Philadelphia to help us this week. As you are probably aware, our farm has friends all over the colonies."

Matt reached out to shake hands with the younger son, and then the older. He made sure his handshake was hard and firm. Paul Payne was about Matt's height. Levi was two inches taller. "Hope you work hard, Mr. Miller. They need all the help they can get," Levi said. "Do they pay well?"

Matt had despised these kinds of men all his life. They were always looking for a fight. Matt thought better of it at first, but succumbed to the temptation to torment the bull—a little. "I work hard, Mr. Payne," Matt replied. "The Taylors have an exceptional farm known all the way up in Pennsylvania. It's the least I can do for the opportunity they give me to get away." He paused for a moment, and just as it looked like Levi was ready to say something, began again. "As for pay," he said, "I work for free. I already have more money than any man requires. I hope to get a few more of my associates down next year to donate their time. We could all use the fresh air." He paused to let it sink in and then looked Levi straight in the eye and said, "Don't you agree, my friend?"

Matt saw the rage grow in the man's eyes and knew that he had gone too far. "Mr. Miller," Levi said coldly, "let there be no mistake. We are not friends." He clenched his fists.

"Levi!" said Nathan. "You'll not fight in here."

Levi stared intensely at Matt. Matt could tell he was waiting for another word to start swinging, but he didn't bite; he remained silent and calm. He slowly shifted his back foot for better balance. Will put his hand against Matt's chest and looked back at Levi. "It would serve neither of us to cause a disturbance," he said. "We can continue another time."

"Wager on that," Levi replied. He had venom in his voice.

"For now, we'll be on our way and you can enjoy your meal," Will replied.

"My new friend should journey back North where he belongs," Levi said, "while he can still walk."

Matt started to reply, but Will raised his hand, making him step back. "Good day, gentlemen," he said. "We have other engagements."

Will let loose on Matt once they were on the street. "Are you mad?" he exclaimed. "It's bad enough many already think we're enemies!"

"Acting like a lamb would only have made him madder," Matt replied. "I'll never see that man again."

"Ah! Levi is an idiot," Will said finally. "It was my boasting about how many friends our farm has."

"The other son seems fine."

"Paul's a gentleman. He was to be married to Kathryn."

"Your sister?" Matt exclaimed. "I can't think of anything worse than your two families together."

"On the contrary, we were close to healing a years-old rift and becoming partners," Will said. "Paul was devastated when Kathryn died."

# CHAPTER 10.

## FARMHAND

---

"Do you go to this King's Tavern often?" Matt asked as they drove away in the buggy.

"Ofttimes, but I'd not have been there this day save for you," Will replied. "The horse people eat there. It's either competitors or patrons; I don't know why."

"Are there other horse farms?"

"Two within a day's ride," Will said, "though we don't have such a singular relation with the others. Many have higher numbers and cheaper labor. We survive on quality."

"Because of Shadow."

"I've exaggerated our dilemma. 'Tis not only Shadow. The stock we've selected rivals that even of the Browne farm. Our pairings have produced animals of the finest quality."

"Was it true what Levi said about your business situation?"

"Half, perhaps," Will replied. "Nathan isn't even our greatest competitor."

"This is all about not having slaves?"

"Probably." Will paused. "They are cheap and skilled labor. Let's see what men Uncle brings out to hay tomorrow." He paused again. "You know," he added, joking, "a drunkard might find a fellow or two among them."

"Funny," Matt said. He shook his head no. "I have to go home."

"You must wait 'til your ring sells."

"Forgot about that."

"'Tis settled, then. You're hired."

"I don't know a thing about farming," Matt replied.

Will repeated Matt's phrase back to him. "What else do you have to do?"

"I'm not actually going to work for free, though," Matt said.

**********

They pulled up to a good-sized white farmhouse that sat back away from the road. There were three carriages tucked next to the farmhouse and then a narrow pasture with two horses grazing. "We arrived before Father," Will said. It wasn't long before they saw the older man's wagon coming down the street. Will shook the reins and they headed to meet his family.

"How does it, gentlemen?" Thomas said, waving. Both the younger Taylor boys waved, along with the mother. Grace smiled politely. Their wagon pulled up alongside the carriage.

"Old Man Hancock has bought the twins for one hundred and fifteen pounds," said Will.

"'Tis *Mr.* Hancock," his mother proclaimed.

"Sorry, Mother. *Mr.* Hancock has bought the twins for one hundred and fifteen pounds. He'll be out to pick them up tomorrow."

"One hundred and fifteen pounds is an excellent price," Thomas said.

"So you are aware, Father, Mr. Miller knew of these horses and their sire," he said, looking at Matt. "He reminded Mr. Hancock of their ability to *shimmer red* in the sunlight. That seemed to close the sale."

Matt was chagrined to see the entire Taylor family now staring down at the twins. They tilted their heads as if in a trance. Matt had to keep from laughing. "They do shimmer!" Jonathan exclaimed.

The elder Thomas looked at Matt and said, "We'll settle once they're sold."

Matt acknowledged Thomas with a nod. "Thank you."

"Oh, and one other thing," Will said. "I've hired Mr. Miller for a fortnight to help with the hay."

"Are you not required to return to Philadelphia, Mr. Miller?" Thomas asked.

"I've arranged for business with the silversmith," Matt said. "It may take a few weeks."

"We'll discuss terms when you return," Thomas said. Matt nodded.

"I must visit my apartment and fetch my clothes," Will announced. "We'll be out to supper after."

"'Tis at four o'clock," Mary said to her son. "Don't make your guest late."

"You have my pledge," Will said as he started the horses down the road.

They had driven only a short distance when he turned to Matt and said, "Let's visit some fellows."

"Fine with me," Matt replied, despite the fact that his head was pounding and he was getting very tired. Matt had hoped they would get the clothes and go right out to the farm, but Will was intent on making the most of his

Sunday. Will pulled the buggy in front of a sign for the Gold Lion Inn. They got down and Will led him past people in the courtyard and then around to the back.

A brawny man saw them as they approached and called out, "Will! How does it?" He looked like a human version of a walrus.

Will shook his hand vigorously. "Very well, Henry, very well. And you?"

"Couldn't be better." Henry turned his walrus gaze to Matt.

"This is Matthew Miller," Will said. "He's staying with our family while he does business in Richmond."

"Henry Boyd," he replied boisterously, reaching out to shake Matt's hand. It was like shaking the hand of a giant. "What business are you in, Mr. Miller?"

Matt was going to say science, but thought better of it. "I'm an apothecary."

A broad grin parted the walrus beard on the man's face. "You've dealings with Benjamin Scott, then," he said, excited. "He's sitting out front."

Will interrupted. "He has business with another, Henry, and probably doesn't want to share the particulars."

Henry put his big arm around Matt's shoulder. "I didn't intend to pry. You keep your secrets!" He slapped Matt on the back and almost knocked him off his feet. "Have a seat, young lads. Rebekah will bring cups of tea. You drinking anything else, Mr. Miller?" Matt shook his head and the man was gone.

"My fellows are there," Will said, pointing to a table of four men looking to be around Will's age. The men were deep in discussion and hadn't noticed them approach. "Gentlemen, make room," Will declared.

They all looked up, and one said, "What's that fellow's name?"

"Hilarious, Robert," Will replied sarcastically.

"Where've you been?" a light-haired man asked.

"You know Samuel!" Will explained. "I hardly have time to piss. This is Matthew Miller from Philadelphia. He has lodgings with us while he completes business in Richmond."

"Have a seat, then," said Robert. "Thank the Lord for the Sabbath! James was telling us about courting Selah Hammond."

"I would not break Miss Hammond's confidence," James replied.

"We are your fellows," Robert pleaded in a high voice. "Fill our empty lives with your tales of romance." He turned his back to the table, wrapped his arms around himself, pantomimed a kissing couple and then said in a fake woman's voice, "Oh, Mr. Montgomery, I love you." The whole table roared in laughter. "What of you, Will?" Robert asked. "Has the Prince of Horses turned his attention to any Richmond ladies?"

"Sadly, no," Will said. "I did take Mr. Miller to meet Ciara at King's Tavern."

A cheer went up. "Sweet Ciara!" somebody yelled. "You didn't leave him alone with Ciara?"

"No, but 'twas bad enough," Will said. "He left her two shillings."

"'Tis worth the silver," someone proclaimed. "I'd pay all my silver to look upon her." They laughed.

"I was bewitched," Matt said, laughing. "She cast some spell on my brain."

"Brains aren't where she casts the spell on me!" someone else said. The whole table laughed again. The bawdiness of the humor surprised Matt. He had never expected colonial men to be so like his own friends.

**********

After a couple of hours, the young men had exhausted almost every topic and eventually started getting up from the table. Will had been deep in the conversation from start to finish. When he finally checked the grandfather clock in the corner, he said, "The time!" Almost immediately, he and Matt were saying their goodbyes, rushing out of the inn, and hopping into the buggy. "We must collect my clothes, and then we're off!" Will exclaimed.

They sped down the road to a small house. Will jumped from the buggy and raced inside with Matt one step behind. The bag on Will's bed was partially packed, and he pulled more items from the dresser and a brand-new hat from the wardrobe. "Try this," he said. "It was a gift. It's too small." Matt set the tri-cornered hat on his head and found that it was quite comfortable. In the pictures, these hats always looked awkward and bulky, but now that he had one on his head, he saw the attraction. Will looked up from his bag and said, "A fine Richmond gentleman. It's yours."

"Who wears these?" Matt asked, looking at his reflection in the window.

"Tricorns?" Will asked as he looked into his bag. "Everyone I know," he murmured, distracted. "Do they not wear them in Philadelphia?"

"Yes," Matt replied hesitantly. Every picture of Ben Franklin he had ever seen had him in one of these hats. "I just don't wear them."

"That one's too nice for the fields," Will said.

Matt stepped back in front of the windowpane to stare at his reflection. He did look the part of a colonial gentleman. Will finished packing and they ran back out to the buggy and sped toward the farm, trying to make up for lost time.

"How long will we bring in hay?" Matt asked.

"A fortnight or so," Will replied. "'Tis not only hay. There's corn and oats, and Uncle also has tobacco."

"You grow tobacco?" Matt asked, surprised.

"Uncle has some conceit we should make money for purchasing stock."

"Is there money in tobacco?"

"Depends on supply. Much gets exported to England. As of late it seems every farm is growing leaf."

"How long will *you* work in the fields?" Matt asked.

"Until the hay's done."

"What about your apprenticeship?"

"It's agreed that I'd be gone for planting and harvest. I desire some respite from Samuel Smith. He works me like a slave."

# CHAPTER 11.

## DAMN DOG

---

It took them about thirty minutes to get to the Taylors' farm. It was much busier than when they had left in the morning. Matt counted ten workers between the road and the farmhouse. David had come alive and was barking orders to the men. He waved to Matt and Will as they arrived and called out, "You're late."

"Your humble servant," Will called back.

Thomas waved and then motioned to Jeb on the other side of the barn. "Finish up and get your brother and sister. It's supper soon."

Will tossed his bag in front of the house and they continued around to park alongside the back of the barn. He unhooked the horse, pulled him to the corral next to the barn, grabbed a bucket of feed mix from a bin, and put it in the horse's trough.

Grace was in the barn, drying a horse with a cloth. Will called to her. "Hoy there, sister. It's almost supper. What's the delay?"

"You're late," she called back. "I must finish with Joshua." There was humor in her voice.

"I'll let Mother know," Will replied. "Try to be prompt," he added, obviously pleased with his ability to torment her.

Scout came bounding up to greet Will, who ran his hands up and down the dog's fur, saying, "Good boy!" But when Matt reached down, Scout refused him with a low growl. Scout bounced along next to Will as they headed to the house and followed them up to the porch, where he waited while the men went inside. Will returned from the kitchen with a big bone for Scout and the dog trotted away happily.

"I'm his favorite," Will said, winking.

Mary was inside cooking. "Are you hungry?" she asked.

"We've not eaten since church," Will replied. "I'm starved."

"We need two chairs from the porch," she said.

"I can get them," Matt replied. He made his way out to the rear porch, where extra chairs were stacked under an overhang. The grounds on this end of the house were dense with trees, but Matt could see through a clearing to another white farmhouse that joined a group of grey shacks. Smoke streamed from an open fire. Matt looked for signs of activity, but no one seemed to be around even to tend the fire. He looked again at the shacks and it occurred to him that this was probably where they housed the temporary workers. He stood enjoying the silence for a moment, then turned around, grabbed the chairs, and walked back into the house.

Thomas was there when he entered. "I'm very pleased my son convinced you to help us," he proclaimed.

"I have time," Matt explained. "It would be good to get in a day or two of honest work."

"You'll get that," Thomas said. "We pay three shillings per day, plus boarding."

"Sounds reasonable," Matt replied. "I do have one request, though. I want to continue to sleep in the barn if that's okay."

"Guess you city boys aren't used to sleeping in camps," Thomas replied. "You can use the barn as long as there's space. We may fill it."

"The dog and I will find another place if we have to," Matt replied.

Will was walking down the steps in more casual clothing. He had Matt's pack in his hand. "I imagine this is yours," he said, hefting it to him. "Do you carry bricks in there?"

"I like to be prepared," Matt replied.

"The label says 'twas made in China," Will said. "I'm amazed that such workmanship comes from the Orient. Seems to be neither wool nor cotton."

"Not sure what it is," Matt said. "Maybe a special type of heavy silk?" He really didn't know. The cloth was some kind of high-tech nylon, thick like burlap, but light and waterproof.

Grace was coming into the house. The front of her dress was wet and she looked troubled. "Joshua is worse than yesterday," she said. "The medicine isn't working."

"I have seen this corruption before," her father replied. "It may spread to other horses. Keep him in his pen."

"His coat is falling out in large patches. It will be a long time before he can take a saddle."

"Give it time," Thomas said.

"You've said this for weeks," she said, her voice cracking. "It's not getting better, but worse."

"'Twill clear," her father said sternly. "We'll speak of this no further."

Matt could see the scorn grow in the young woman's face. Her voice quaked with rage. "I'll not mention it, Father, but this too will not be solved by not speaking of it."

"Enough!"

She glared at him and then walked upstairs to her room, shaking her head. Her older brother looked at Matt as if to say, "I told you so."

Mary was beginning to set dinner out on the table. Matt arranged the chairs to match the plate settings. There were now only seven.

"Will David and Faith not join us?" Matt asked Mary.

"They're cooking for the men. There'll be a feast tonight to celebrate the harvest."

"A feast?" Matt asked.

"We're known for using our workers well," she explained. "Men from the nearby towns seek David out during harvest time. God has blessed us this year with many who have been to past harvests."

Grace was coming down the steps in a simple blue and white country dress with a ribbon tied around the waist and tan leather slippers. Matt was talking with Will about the next day when he look up and saw her. He had to force himself to turn back to Will, realizing that he had stopped midsentence. There was a giant grin on Will's face. "You were asking where all the corn would be held," Will reminded him.

The elder Thomas called out, "Dinnertime." The younger boys immediately sat at the table, followed by the rest of the family. Mary put a bowl of cut carrots on the table and joined them. "Let's pray," she said. She glanced at Matt, thought better of it, and looked at Will. "Son, would you say the prayer?"

"Certainly, Mother." Will began to pray, looking out over the table.

*May the peace and blessing of God descend upon us*
*As we receive of His bounty,*
*And may our hearts be filled with love*
*For one another.*

He paused for a moment as if he'd end, but then continued,

*Bless, O Lord, Almighty God, this farm;*
*May health and purity, goodness and meekness, and every virtue reign here,*
*May all those who dwell here be filled with faithfulness*
*To Thy law, and with thanksgiving to God, the Father, the Son, and the Holy Spirit,*
*May this blessing remain on this farm and all who dwell here,*
*Through Christ our Lord,*
*Amen.*

"That was beautiful," Mary said, beaming with pride.

"Suggested by the reverend," Will explained. "To enter the fields and plant so soon after Kathryn's death; 'tis a testament to God's healing power." The table went quiet. "A time for joy!" Will exclaimed in response to their solemn stares. "I know that Kathryn is looking down upon us from heaven to aid us in our health and prosperity."

"I agree," his father declared. "Let us take a joyous meal in her honor!"

Plates of roasted pork with strong vinegar sauce and bowls of boiled potatoes and carrots were passed around with a comfortable enthusiasm. "What time will we rise tomorrow?" Jonathan asked as he spooned out carrots.

"Uncle will wake us at sunrise," Thomas replied. "We'll eat a hearty meal and then we must labor until dark."

Jeb groaned. "Sunrise?"

"Sunrise!" his father repeated. "The Lord doesn't provide us with these gifts that we should leave them in the field."

Jeb groaned again.

"Tough work for a city boy," Grace said to Matt.

"Not exactly," Matt replied. "I had lots of manual jobs before I became an apothecary." It was the truth. Matt's parents were rarely around when he was a teenager, so he had occupied his time at anything that paid a wage.

"You'll do fine, Mr. Miller," Thomas said reassuringly. He turned to his son. "You'll want to keep your gloves on. I can't imagine the quill has prepared you for harvest."

"That reminds me," Will said. "Nathan wished you well."

"It's been a long time," Thomas said. "I regret not being able to visit him more often."

"They could make an effort to come to us," Mary said. "I've invited Sara often to Sunday dinner and they've always had other engagements."

"I'm in no hurry to see them," Grace mumbled.

"I know as well as you," Will said in response to his sister, "but Nathan's greetings to Father were heartfelt."

"Naught justifies the tolerance he shows for his son," Grace said coldly.

Will turned to Matt to explain. "Grace has little admiration for Levi." He gave a subtle wink to Matt.

"Levi is cruel to his animals and his slaves," Grace said. "There's no excuse for the things I have seen."

"He cares for his property as he sees fit," her father replied.

"A wretched man!" Grace said firmly.

"Grace!" her mother said. "'Tis not Christian to judge him so harshly."

"Enough! 'Tis harvest time and we should celebrate," Thomas said. "God willing, there's much labor ahead of us."

"Has it been a good season?" Matt asked.

"We'll not know until we're finished, and there'll be much time in the Lord's service before then," Thomas said. "I don't think I've ever seen the corn so tall."

"Father loves the corn," Jonathan proclaimed. "Sometimes he walks forever in the fields."

"I've been with Father often in the corn," Jeb added. "I have never seen him so happy as when he looks out over the cornfield. It's his favorite thing."

"It's not my favorite thing," Thomas said, "but I do enjoy watching it grow." Matt saw his eyes grow misty. "Many people look for proof of God, but the corn reminds me of his presence, every day."

"How?" Matt asked.

"As you sow, so shall you reap," the father explained. "I see the enterprises of men in the corn."

"Enterprises?" Matt asked.

"Men must all plant seeds in their lives," Thomas said. "They are required to care for the seeds as they grow in the spring and summer months." He smiled and exclaimed, "Ha! Even then, as the plants are grown, they aren't finished. They must harvest in autumn or everything will be lost. Men must trust in Him that the seeds will sprout, that there'll be enough rain, and that a hailstorm won't smash all the plants to the ground."

"So growing corn is like the things that men try to accomplish," Matt said.

"Think of your experience," Thomas continued. "Do you know men who expect riches despite the fact that they have not planted seeds?" Thomas paused, measuring his words. "I have encountered many godly men who stand above the untilled ground praying, but without the investment in the seed and the labor for planting, the corn doesn't grow. There's still the risk that the plants are destroyed, but you shall never learn if the seeds have not been planted."

Thomas stopped, embarrassed. "But sometimes it's as my son said. I enjoy watching the corn grow."

They finished dinner and got up to clear the table. Thomas motioned to his eldest son and Matt. "Let's bring the horses in."

As they walked to the pasture, Will said, "We can't forget that Old Man Hancock will be out to buy the twins tomorrow. I hope Grace can get them brushed before he arrives. I know he'll want to show his friends."

"Your sister never lets the horses off the farm until they shine," Thomas said, smiling proudly.

Closer to the pasture, Will called out for Scout, who strode around the corner of the barn. Matt waved Will quiet to see if the dog might listen to him. "Scout," he said as he pointed to the corral. "In!" To Matt's satisfaction, the dog rushed into action, sprinting to the edge of the fence to begin moving the animals into the corral. This time they only needed to stay out of the way and watch. Scout hugged the fence, keeping the horses out of the corners, and soon had them all rounded up and in the corral. When the gate was shut, they headed back to the house.

"David will welcome the men with a prayer," Thomas said. "We made him pledge not to serve the ale until we returned."

"You let them drink?" Matt asked. It seemed surprising after hearing the disgust in Grace's voice when she said the word *drunkard*.

"It's not what I would choose," Thomas confessed, "but Uncle insists that we provide the men with drink for the evening. We've debated this many years, but the reality is that some won't come to stay the week without drink. Others would bring their own bottles." He shrugged. "If we provide, they are less likely to get so drunk that they can't wake up."

By the time they reached the common, the sun was starting to go down and the trees threw long shadows along the path. There was a crackling fire burning in a pit next to where the food had been cooking. The night was comfortably warm, so most of the men weren't sitting near the fire, but were eating, talking, and laughing at tables scattered here and there around the common. Matt counted sixteen men as David and Faith moved among them filling and refilling plates and cups. David saw them and called out, "Men, please welcome Mr. Thomas Taylor."

Thomas stepped onto a crate in the center of the common. It was now dark enough that the firelight shone on his face. The workers quieted. "Men," he said. "Many thanks for joining us. As you know, we have a reputation for producing the finest horses in Virginia, and this would not be possible without your contributions. We'll be paying you a fair day's wage and feeding you like kings. Since I don't want any of you to become fat from all these good victuals"—he grabbed his belly and the men laughed—"we'll be working you very hard."

Thomas smiled at the good-natured jeers this got. "There's much ahead of us, but 'tis not all bad. We'll divide into teams tomorrow. There'll be a prize to any group that exceeds their standard."

A cheer went up.

"We'll have a prayer first and then we'll break out the ale," Thomas said. "Drink up, but sleep well. The prizes await."

A cheer went up again.

David took Thomas's place on the crate with a Bible in his hand. "Men!" he said. "You're working for the finest Christian family in Virginia, and they expect you to act accordingly. Let us pray!"

*Almighty Lord God,*
*You keep on giving abundance to men in the dew of heaven,*
*And food out of the richness of the soil.*
*We give thanks to Your most gracious majesty,*
*For the fruits of the field which we have gathered.*
*We beg of You, in Your mercy, to bless our harvest,*
*Which we have received from Your generosity.*
*Preserve it, and keep it from all harm.*
*Amen.*

Thomas waved. "Drink up tonight! We begin tomorrow at sunrise." He reached down and lifted a barrel of ale onto the table with a thump. Thomas nodded to his son and Matt. "Stay and drink. Get to know them." He led the rest of his family back to the house as David filled tin cups with ale.

Three cups later, it was almost ten o'clock and David was breaking up the party. Matt grabbed his pack from the house and headed to the barn with a lantern. The bench he was using as a bed was set up from the previous night and his jacket was hanging from a nail on a pole next to the bed. He pulled his phone from his jacket, checked that the door was shut, and turned music on while he spread out his blankets.

As he arranged his bed, he heard Scout at the door and slid it open. The dog trotted in and jumped up, taking most of the covers. Matt reached down to grab the blanket, but Scout growled and he yanked his hand away. He pulled what remained of the blanket around his body and tried to sleep as best he could, thinking, "Damn dog."

# CHAPTER 12.

# NATURE'S BOUNTY

---

Morning felt like it came instantly as Matt was roused by David's pounding on the barn door. "Mr. Miller, wake up." He opened the door and repeated, "Mr. Miller, wake up."

"I'm awake," Matt said in a muffled voice. "I'm awake." The grey morning light was streaming into the open windows. Matt sat up and waited for his head to stop throbbing.

"Breakfast is almost ready," David called. He was now somewhere far away. Matt looked over at the dog, hoping for a companion to share his plight, but the animal was snoring again after having opened one eye. Matt pulled on a pair of pants and a tee shirt, laced up his shoes, and headed to the privy. The seat was cold, and it was dark inside, and the Taylors were waiting for him, so he hurried. People always said everything moves fast in modern times, but so far in 1762, there was barely time to sleep.

Matt was beginning to see the full picture of the immediacy of the lives of colonial farmers. This was harvest time, and there were approximately two weeks to get the hay under cover. If they didn't work hard in these two weeks, the hay would be of lower quality or ruined entirely. If that occurred, the farm would be short on feed and would need to buy it from other farmers. There was no one to fall back on if they ran out of money or failed; the family alone would suffer the consequences. Based on this, there was no compromise and it was all-out work from dawn to dusk.

After he finished in the privy, Matt went to the well for a drink and to fill the washbasin. The water was cold and refreshing and he was getting used to its iron taste. There was a table there, so rather than move the basin back into the barn as he had before, he got his shaving kit and washed and shaved beside the well. The cold water helped shake him out of his morning stupor. Matt wet his toothbrush and squeezed mint-lime toothpaste onto it. He looked down at his brand-new toothpaste and dental floss and realized that if he was stuck in 1762, he might be looking at his lifetime supply of dental products. If he remembered correctly, most people in colonial times did little more than wipe their teeth with a wet rag.

"Mr. Miller, what are you doing to your mouth?"

The youngest boy, Jonathan, had snuck up behind him. Matt looked down. "I'm brushing my teeth," he said around a mouthful of suds.

The boy was staring at him strangely. "Uncle said to make sure you were awake. Breakfast will start soon."

Matt spit the suds out onto the ground.

"Is that soap?" the boy asked.

"It's toothpaste," Matt replied. He was brushing his front teeth now.

"Does it hurt?"

"It feels good. It's refreshing to brush your teeth."

"It looks like soap. I hate when I get soap in my mouth."

"This kind doesn't taste bad. It makes your breath fresh." He gave Jonathan an exaggerated wink. "And the ladies love a man with fresh breath."

"Can I taste?"

"Shouldn't we be going to breakfast?"

"It's not out yet," the boy said. "Let me taste."

"Fine," Matt said. "Give me your finger." Matt was rinsing his toothbrush out in the basin. He grabbed the toothpaste and the boy's hand and squeezed some out onto his fingertip. Jonathan hesitated for a moment, then sucked the toothpaste off his finger with a smack.

"It tastes good."

"You're not supposed to swallow it," Matt said.

"Can I try the brush?"

"Where I come from, you're not supposed to share your toothbrush," Matt said, trying his best to be discouraging.

"Why?"

"Lots of reasons that I don't want to go into right now."

"What reasons?"

Matt's head hurt too much to give a long explanation. "I'm hungry," he proclaimed. "You can try the brush some other time."

"Capital," Jonathan said. "I smell breakfast." He was gone as fast as he had appeared.

By the time Matt arrived in the common, most of the men were already in line for scrambled eggs, bacon, and potatoes. Some held out tin cups for black coffee from a metal pot. Matt grabbed a plate and sat at one of the benches with a few of the other men. He ate there, mostly staying silent to listen to their conversation and enjoy the food. He stopped for a moment after his first couple of bites, thinking that it was the most delicious bacon-and-egg breakfast he had tasted in a long time. Even the coffee, bitter and full of loose grounds, had an appeal that was surprisingly new, plus the caffeine helped to dull the throbbing in his head.

Once everyone had eaten—or eaten seconds, depending on their appetites—David walked around and gave out assignments. Matt would be on a scythe crew with Will, a large man named Charles Finley, and two others. David warned Will to make sure Matt knew what he was doing before getting too far into the field. From what Matt could discern, they would be mowing in parallel and clearing the field in a circular fashion. They would cut the hay first thing in the morning to take advantage of the dew, and the hay would get harder to cut as it dried in the sun.

The men climbed into the wagon and headed to the field with Will driving. After some time they turned off the main road and headed to a gap in the trees that opened onto a massive clearing. "We got the big one," Will declared. "My father and Nathan cleared this when I was a baby. It was only the two of them and a few horses."

The men dropped to the ground and began pulling scythes from the back of the wagon, testing them until they each found one that was appropriate. Once they'd placed their selections on the ground away from the wagon, each went for a leather holster and a sharpening stone. The belt of the holster was buckled around the waist and then the holster was filled with water to keep the stone wet. After watching how it was done, Matt strapped on a holster and filled it with water. He picked out a stone from the box and dropped it in the holster. The water splashed up out of the sheath and soaked his leg; he had filled it too high.

Matt walked to the wagon with his wet leg, picked out one of the scythes and hefted it in his hands in the way he'd seen the others do. He placed it on the ground away from the wagon. Charles, who had been watching him, walked over, picked out a scythe and brought it to Matt. "Try this," he said. "The one you chose is too short." Matt had expected his voice to be gruff, based on his large size, but Charles' words were measured and calm.

"Never used a scythe," Matt replied.

"I want that prize today," Charles said. He spent some time showing Matt how to use the scythe.

**********

It took about an hour to do the outer circle of the field, after which they took a break to drink water and refill their holsters. Ten minutes later, they returned to mowing, with everyone shifted five lanes to the center. Matt's lack of skill became evident as the sun dried the field and it became necessary to slice the grass rather than the hacking that was possible when it was wet. When they shifted over for the third and final swath, they found that it was only four men wide, and since Matt was the slowest, Will instructed him to grab a small scythe and clean up the edges of the field.

By the time Thomas rolled up in the lunch wagon, they had finished mowing the entire field and stood at the wagon, resting and drinking water from tin cups. "All done?" he asked. His son nodded and Thomas exclaimed, "A grand field to cut in one morning!" He stepped aside and quietly talked to his son, trying to plan the remainder of the day. While this was going on, Charles and the two other men, who Matt now knew to be Elias and Zachariah, were getting their plates from Mary. Thomas had moved stools from the back of the wagon out onto the ground. Matt sat as he ate his cooked pork, squash, and sugared pears.

"Baked yet?" Charles asked.

"I'm a little sore," Matt replied. "I kept hitting the ground."

"I was very sore the first time I used the blade," Charles explained. "It grows easier."

"I'm learning," Matt said.

"God willing," Charles replied seriously. "There's another prize tomorrow."

Mary collected the plates, and then returned with a tin of water that she handed to Will to place in the back of the wagon they would take to the next field. When he finished eating, Will came over to their team. "The weather should hold," he proclaimed. "We can wait until tomorrow morning before we mow again, so we'll spend the rest of the afternoon in the tobacco field."

The men piled into the wagon and Will drove them to the tobacco field. Based on Will's description, Matt had expected a small experimental field, but it was nothing of the sort, with plants that covered a large area. The dark green tobacco leaves came to right below Matt's chest. The bottom leaves on most of the plants had started to yellow and a number of the smaller plants had seed stalks growing from their tops. Matt learned that the seed stalks on the older plants had been snapped off in a process called topping, to concentrate the oils of the plant in the lower leaves.

The men jumped to the ground and grabbed their hatchets. By late afternoon, they had filled six wagons with tobacco stalks and were on their way to join the other groups at the farm for dinner. They jeered as they rode past one worker, who shouted, "Lazy buggers."

Many had been in the hayfields all day and were covered with grey-green chaff. As they jumped from the wagons, they took time to shake the dust from their hair and sweep the chaff from their bodies with brooms. Sweat glued the chaff to their arms, so some headed to the well to fill buckets and wash.

By the time everyone returned from their break, Faith and David were serving dinner in the common, and the men were already talking about heading back into the fields. It seemed to Matt that his first day of farm work might never end.

# CHAPTER 13.

# IBUPROFEN

---

Matt was relieved to find that the evening was calmer and lacked the urgency of the daytime harvest. They took their time walking through the field with big forks, turning the hay so the unexposed side would dry in the morning sun. The men talked as they worked, about Richmond, their homes, their families, and of course, women. Elias and Zachariah seemed to be friends, and talked often of the girls they knew in town. Zachariah's brother was to be married to one of the Richmond farm girls and there was controversy over her small dowry because he was only a blacksmith's apprentice. Many townspeople were sure the couple would have a financial struggle from the very start.

Matt had been lost in his thoughts about Richmond, when unexpectedly, Elias said, "Mr. Miller, where did you say you were from?"

"Philadelphia," Matt replied. Two things about answering questions about his life started to bother Matt. One was that he had been telling so many half-truths that it was hard to keep track, and the other was that he had been so convincing and compelling in narrating his made-up life that he was starting to believe it himself.

"That's a great city," Zachariah said. "I heard the ladies are friendlier than in Virginia."

"I'm no expert on Philadelphia ladies," Matt said, "or any, for that matter." He smiled, thinking that it was his most truthful statement in days.

"Can anyone predict ladies?" Charles said in a booming voice from three rows over.

"I think Philadelphia ladies are more conservative than most," Matt explained. "We always think that the ladies in New York are friendlier."

"They're always more friendly somewhere else," Will said, chuckling.

"The grass is always greener on the other side of the fence," Matt replied.

"What does grass have to do with Philadelphia ladies?" Elias asked.

"It's Philadelphia expression," Matt said. "It means that you always look at another man's life and assume it's better than yours." In the silence that followed, Matt couldn't resist blurting out, "Ladies! You can't live with 'em, and you can't live without 'em." He felt shameless, but he might as well have been Plato or Aristotle based on their reaction.

"There is no statement truer than that!" Will declared.

They finished the field quickly and were soon on their way home in the wagon. Thomas greeted the men as they arrived. Like the night before, as soon as he saw Will and Matt, he asked them to bring the horses in. They went immediately to the pasture.

"Scout!" Will called. The dog came running around the corner, trotting to Will and Matt as they walked. He stared at Matt and gave a low growl.

The dog took his place beside Will and made sure to glance suspiciously at Matt as they walked, which Matt tried his best to ignore. Once they reached the pasture, they were able to get the horses into the corral quickly and were soon on their way back to the common. They took a shortcut through the barn and saw that Grace was drying a horse. Will stepped to the fence and said, "How does Joshua?"

"Worse," she replied.

"Father believes he'll heal."

"Father might be wrong."

"Try to finish. We have evening prayer."

"I may not be there."

"You know how Father gets when you're not there for prayer," Will cautioned.

"If Father is so vexed, maybe he should care for this horse."

"Come to prayers."

"If you give me some peace, I can finish," Grace replied.

Will waved to Matt and led him out of the barn. "Joshua looks bad," he said. "Father is ready to put him down. He fears he'll spread the infection to the other horses."

"How long before he decides?" Matt asked.

"Don't know…days, weeks. It depends on his trepidation."

Most of the men were sitting at the wooden tables when they reached the common. Thomas and David walked about, thanking and congratulating them on their first day. Thomas uncovered the ale that had been on the center table and said, "Drink up, men. You've worked hard today." He left soon after with his family in tow.

Matt and Charles sat alone at the table closest to the house as the Taylor family passed. As soon as they were out of earshot, Charles said, "She's as beautiful as her sister." He looked around, making sure no one was nearby.

Matt could see Will far away, talking with David and pointing in the direction of one of the fields. He wasn't sure he wanted to enter a conversation about Grace with a man he didn't know that well, but he was intrigued to learn something about Kathryn. "Did you know Kathryn?"

"Sure," Charles replied. "Half the men in Richmond wanted to marry her."

"She was engaged to Paul Payne," Matt said.

"Paul's a capital fellow," Charles said. "His brother Levi, well…"

"I met Levi," Matt replied. "He didn't seem to be a friend of the Taylors."

"He's not a friend of much besides showing his father's money," Charles said. "That and fighting."

"How did Kathryn die?"

"She took a hard fall off a horse, and she was bleeding inside. 'Twas at the Payne farm."

"She fell off a horse at the Payne farm?"

"That's what they say," Charles replied. "Ladies don't belong on horseback."

"Do you ever see ladies riding on horses in Richmond?"

"Never in town. Hither, though," Charles said. "The Taylor ladies were often seen sitting astride horses, and some pretend to have seen them in men's breeches."

"Men's breeches?" Matt tried to sound as scandalized as he could, though he was hiding a smile.

"The Taylor ladies took much license in the way they carried on."

"It sounds like," Matt affirmed.

"Mr. Taylor has put an end to that nonsense," Charles said. "Grace isn't allowed to ride. I don't know how she'd have the time with all the men in Richmond wanting her attention."

"She has a lot of men wanting to marry her?"

"They'd never allow her to be married to a simple farmhand," Charles replied. "Some pretend Levi fancied Grace. He'd be wealthy enough."

Their conversation was interrupted as Will approached with three tin cups of ale balanced in his hands. "Ale for my fellows," he declared. They sat together, drinking and talking about the day. Matt tried his best to remain upbeat and energetic, but as the conversation wore on, he became mostly interested in going to bed. They made a mutual decision to call it a night after only two cups of ale. Charles said good night and headed to his shack and Matt walked back towards the house with Will.

"I'm sore already," Will said, rubbing his shoulder. "I'll have trouble getting out of bed."

"Come to the barn," Matt replied. "I have medicine that will make you feel better."

"Are you selling me a magic tonic?"

"Not even close," Matt said. "This is a very expensive and rare medicine." They walked to the barn with a lantern. Matt opened the door and went inside for his pack. He pulled out his shaving kit to find the bottle of Advil. *The world's supply of ibuprofen.*

Matt pulled out two for Will and two for himself and grabbed his cup from its hook on the wall. "Let's get water," he said. At the well, Matt held the tablets up for

Will to see. "You're supposed to put these in your mouth one at a time, drink water, and swallow," he said.

"Without chewing?"

"Without chewing. The medicine tastes bitter if you chew."

"Makes sense."

"Watch me," Matt said. He swallowed the two tablets and then said, "Your turn."

Will popped a tablet in his mouth and took a gulp of water, tilted his head back, and swallowed as Matt had demonstrated. He coughed and the pill popped out of his mouth covered with spit and landed on the table.

"You need to swallow it," Matt said. "Try again." Will put the wet tablet back in his mouth, filled it with water, and swallowed. This time it went down and he opened his mouth to show Matt it was empty.

"One to go," Matt declared. The second tablet went down easier.

"Now what?"

"You feel less sore within the next half hour," Matt said. "You sleep well, and you feel better in the morning."

"This medicine pledges much," Will replied.

"Let me know whether you think it worked," Matt said knowingly. He considered ibuprofen one of the greatest drugs of the twentieth century. Aside from his academic appreciation of the drug and its discovery, he had often used it after sparring sessions in tae kwon do. As Matt finished with Will, it occurred to him that people here might especially prize ibuprofen since most people did manual labor that was guaranteed to make them tired and sore.

Matt said good night and headed to his barn. He crawled into bed and closed his eyes, but sleep would not come as his conversation with Charles raced through his head. *I'm only a farmhand here.* Matt sat up, reached to his jacket beside the bed, pulled out his phone, and turned it on to listen to music. It wasn't much longer when he heard the familiar scratching at the door. He rearranged the covers so the dog couldn't steal them and got up to open the door.

"Hello, killer," he said. The dog wandered in, jumped on his blankets at the foot of the bed, put his head down, and closed his eyes. "That's probably a good idea," Matt said, but before he could reach his phone to turn it off, it beeped. The dog's ears shot up and his eyes opened wide. Matt picked up the phone and saw a text message on the screen.

"Matthew Miller, are you there?" it read.

"This can't be," Matt said to the dog.

Matt typed, "Who are you?"

"Oak Ridge Laboratories. Can you tell us your exact date and time?"

Matt pulled out his watch, then typed, "August 2, 1762, 9:40 p.m."

"Thanks. Any contact with the others?"

Matt typed, "What others?"

"Time travelers, like you."

Matt typed, "No one else. How did I get here?"

"Reactor accident caused a wormhole. Will text again in exactly twenty-four hours."

"Can you get me home?"

"Soon."

"When?"

There was no reply.

"Dog," Matt declared, "I'm going home."

# CHAPTER 14.

# LASTING IMPRESSIONS

---

Matt slid helplessly into the open pit and tried desperately to hang on by clinging to dirt, rocks, and vines. He watched in horror as the flesh was pulled from his hands by everything he tried to grab. Blood covered his fingers and the wet slipperiness accelerated his descent into the hole. The sides of the hole collapsed and disappeared, and he was falling into open space towards a black abyss that had formed below. He flailed, trying to right his body, hoping he could at least fall to his death feet first.

As he righted himself in his fall, he was able to focus and become calm as he resigned himself to die. The walls were suddenly back and rushing by, coming alive with moving pictures of events and people. Most were shadows, but he recognized a few and many of these were tethered to other shadows by glowing tunnels of light. He passed by as the space shuttle exploded in the blue sky. Connected to another tunnel, he could see shadows of American soldiers with machine guns wandering through a jungle. He saw the Beatles on a platform, the entire stadium screaming as John Lennon swept his elbows back and forth across a keyboard. There were smokestacks as an industrialized America rebuilt Europe devastated by war. These images yanked past him and were distorted by a vision of torches of marching men at Nuremberg.

He looked down at his feet and realized that he was no longer falling into an abyss; there was form now. He was plummeting towards dirt and ground and could see his death approaching. The closer he came to the surface, the more the wind buffeted his body as he fell, and he began to oscillate in the wind and hear booming. *Thwap! Thwap! Thwap! Thwap! Thwap!* It became deafening and he resigned himself to his fate.

\*\*\*\*\*\*\*\*

Matt opened his eyes. He looked up into the darkness. He was no longer falling into the hole. He heard the sound of a fist on wood, *bam, bam, bam*, and then, "Mr. Miller, are you awake?" Then again, "Are you awake?"

"I'm awake," he called back, trying to remember where he was. He sat up and looked at the windows. *I'm in the hay barn.*

Matt stood up slowly in the dark, pulled his pants on, and let himself out of the barn to find the privy. He could hear and see activity on the farm as it began to wake, and the smells of smoke and cooking food were heavy in the air. Despite the ibuprofen, his upper body was sore from swinging, chopping, and spearing. He knew it would've been worse without the drugs. His hands were sore, but he was happy to see that he had no blisters. The throbbing in his head was milder today, but he still needed to reach up and massage his temples. Now that he knew what had brought him here, he understood that the headaches were a side effect of falling through a wormhole.

From the privy, he walked back to the barn and slid the door open. The dog opened his eyes, jumped off the bed, and skirted out, almost tripping Matt in the process. "Where are you going?" Matt called out, but the dog didn't turn back. Matt reached for his phone to see if he had somehow dreamed the messages, but the text trail appeared on the screen. It would not be long before they rescued him and he was back in his own time. He considered it for a moment and decided to leave as good an impression on these farm people as he could. He would work hard the next few days to help them harvest their fields. The idea of being remembered across centuries was fascinating. Even the stunning Grace Taylor might regret not taking the opportunity to get to know him better once he disappeared as mysteriously as he had arrived.

Matt grabbed his pack and the washbasin and with a revitalized vigor walked out to the well. The water seemed colder this morning, and the washcloth was especially coarse as he rubbed it on his face. He felt his chin and noted that it was rough, but he didn't think he had time to shave. He pulled out the toothbrush and the plastic tube of toothpaste.

"Good morning, Mr. Miller," Jonathan said.

Matt spat his toothpaste out onto the ground and said, "How are you doing today, John?"

"'Tis Jonathan." He said it in a way that Matt knew wasn't meant to offend.

"Sorry. How are you doing today, Jonathan?"

The boy was quiet for longer than a moment, and Matt could see he was thinking hard. "I'm fine, though I'm tired from moving the tobacco into the drying house yesterday." He was watching Matt as he was brushing his teeth. "Can I try the toothbrush?"

Matt rinsed his mouth. "Jonathan, I don't think it's a good idea. Sharing toothbrushes spreads germs."

"What are germs?"

"The things people pass around that make you sick. You know how sometimes you get sick from being around someone else who's sick?"

"Sure, Mother warns us all the time," the boy answered. He looked Matt up and down and asked, "Mr. Miller, are you ill?"

The temptation floated through his head, but Matt didn't have it in him to lie. "No, I'm not ill."

"Good," the boy said, excited. "I'm not ill, either." He thought for a moment and said, "Does that mean that I can try the brush?"

"I don't know," Matt said, resigned. The boy looked at him in anticipation.

"You can try it this once," Matt said. He rinsed the brush and shook it out onto the ground, then squeezed out a generous dollop of toothpaste onto the bristles and handed it to Jonathan. The boy took it gently, careful not to lose any of the toothpaste.

"There you are," Matt said. "Brush away."

Jonathan put the brush up to his mouth. "How do I do it?"

"You put it all the way in your mouth and brush back and forth," Matt said. "It should rub on your teeth like this." He used his finger to demonstrate on his own teeth.

Jonathan put the brush in his mouth and moved it around, but still wasn't doing it right. "Let me show you," Matt said. He gently took the brush from the boy. "Open your mouth." Jonathan opened and Matt brushed his back teeth, demonstrating the motion. "Okay, now spit out the suds." Jonathan spat the toothpaste onto the ground as he had seen Matt do. "Give me a big smile," Matt said. Jonathan smiled, and Matt showed him how to brush his front teeth. "Now you try." Matt gave him back the brush. This time Jonathan did a credible job.

"That should do it for now," Matt declared. "Rinse your mouth with water and don't swallow." Jonathan handed the dripping brush back to Matt while he grabbed a cup of water and rinsed his mouth.

"My mouth feels cold," Jonathan said. He whistled as he inhaled quickly.

"Refreshing, right?"

"Refreshing," Jonathan replied as he winked. "And the ladies love a man with fresh breath."

"I was joking about that," Matt said. "That would probably not be something to repeat in front of your mother."

The boy smiled. "We should go," he said. "Breakfast is ready." Jonathan turned around and headed to the common. Matt could hear him whistling as he sucked air through his mouth and then exclaimed, "Refreshing!"

Matt thought that even in the twenty-first century it was hard to beat mint-lime-flavored toothpaste. He cleaned up his pack, took it back into the barn, put it into a storage cabinet and covered it with an old blanket he had found in the barn. Matt entered the common as most of the men were already finishing their food. *Kid made me late.* Matt hurried to fill his plate and sat across from Will, who was nearly finished.

"That medicine you gave me worked," Will proclaimed. "I slept like a log."

"Ibuprofen is good stuff," Matt replied.

"What plant is it from?" Will asked.

Matt thought for a moment to get it right. Most if not all medicines familiar to the American colonists came from plants. Unlike aspirin, which had its origin in natural product chemistry, ibuprofen was a product of man's ingenuity more than a fortuitous discovery. The precursor of aspirin, on the other hand, salicylic acid, came from the bark of the willow tree. Willow tree bark had been used to relieve headaches and pain since before the birth of Christ, but in its natural form was very hard on the stomach. To synthesize aspirin, salicylic acid was modified using acetyl chloride or acetic anhydride to change a hydroxyl group into an ester. Matt could see the synthetic route in his head.

"It's from the willow tree," Matt said. It wasn't exactly true of ibuprofen, but aspirin was close enough.

"The apothecary sells willow bark," Will said, "but this didn't make me ill. I may desire more after today."

"What's today?"

"We're turning the hay. After it dries, we'll gather it into wagons and bring it home."

"That doesn't sound bad," Matt said.

"More lifting," Will replied. "The hay must be out of the fields before the rain."

"Rain?" Matt asked, looking up into a clear blue sky.

"Uncle says rain."

"How does he know?" Matt asked, now wondering what people did before The Weather Channel.

"He can feel it in his bones," Will explained. "We jest, but he's usually correct."

"We better work fast, then," Matt replied as he finished the last of his coffee. He breathed in deeply, taking in the freshness of his surroundings before standing to return his plate.

When they got to the field, Matt was assigned again to the center, but today he kept up easily with the rest of the group. Now he could sharpen his blade as quickly as the others, which helped him swing the scythe more efficiently. They finished mowing the new field and then spread the hay using pitchforks to help it dry. There was a ten-minute break before they proceeded to the north hayfield and turned the hay they'd cut the previous day. As Will had said, this didn't take much skill, but it was backbreaking. They finished turning by midmorning, and were taking their break as the lunch truck rolled up.

Matt was one of the last to receive his meal. Grace placed a big chunk of bread on Matt's plate and he thanked her politely. "You're welcome, Mr. Miller," she said. He had already turned away to go back to his bench when she called, "Are you getting a feel for farming?" Matt turned around to face her. She hadn't spoken directly to him since the first day they'd met. Even when they talked after church, Grace had seemed to be speaking mainly for the benefit of the churchyard.

"I'm getting good with the scythe," Matt answered. "I'm tired from yesterday, though."

"All the men are tired after the first day," she said, "even the experienced ones."

"Some more than others," Matt replied. He raised his hand to his mouth and faked a yawn. Grace smiled and held his gaze for longer than usual, then returned to cutting the bread. Matt stood there watching her hands move.

Seeing him standing there, Mary stepped over with a pensive look on her face and asked, "Is there anything else, Mr. Miller?"

Matt caught himself and answered, "Oh, no, thanks." He gathered his wits and walked away, but not before noticing Grace eyeing him with a sly smile.

After lunch, they shuttled the wagon to the tobacco fields, where the men were able to harvest another third of the field before midafternoon. While they worked, David and Thomas arrived with another empty wagon. They stopped to inspect the full wagons and Matt overheard David say, "I have never seen leaf this green."

"We may make a profit after all," Thomas replied. "God willing."

"God willing," David repeated.

Thomas gathered the crew's attention. "We expect rain within two days," he said, "so we must fit five days of harvest into four. We shall sup in the fields, take shorter rests, and work until sundown. Let us meet this trial so you can return to town for the Sabbath one day early and with a full week's pay."

The sun was low in the sky as they arrived back at the farm. Charles and Thomas moved their wagons to the pavilion and Matt's team joined the others, who were already unloading their wagons. The hay was being stacked under the roof of a pavilion built to protect the harvest from rain and snow. The bales were much easier to unload than to carry out of the fields, so it didn't take long.

Matt wasn't sure how the division of labor developed, but he found himself acting as haystack organizer. He took the bales from the men as they were unloaded and placed them on the stack. If done incorrectly, it would topple when it was about chest high. Matt found that if he did it right, he could stack hay almost up to the bottom of the pavilion roof. He had to pull a bunch of stacked bales to reset them properly and it took him another twenty minutes before he could add more. He lost himself in this and the sun was gone by the time he finished.

He stepped back after he was done and gazed at the tight stack, proud of his handiwork. It wasn't until he was completely done that he noticed that he was alone in the middle of the stacks. He sat there for a time with his two lanterns, enjoying the quiet, then got back on his feet. He was covered in hay dust, so he headed to the horse barn with a lantern in each hand to see if he could find a broom.

Another lantern was burning in the barn, so he expected to see someone, but found it empty. He grabbed a broom and started to sweep the dust from his clothes. His shoes cleaned easily, as did the bottom half of his legs. By holding the head of the broom with his hands, he was able to brush his chest and most of his front, and then he had to contort his body to try to reach his back. The back brushing didn't go well, so he gave up. He grabbed the middle of the broom and started on his legs, but as he was brushing, the broom shifted in his hand and smacked him in the eye, bringing his headache back immediately.

"Bastard!" he said, throwing the broom to the ground.

"Such language," Grace said, peering out from a horse stall. She laughed when she saw Matt standing with the broom at his feet, rubbing his eye. Matt hoped she hadn't witnessed his whole tantrum.

"Are you trying to brush yourself off with a long broom?"

"What's it look like I'm doing?"

"Cursing the floor," Grace replied. "There are hand brooms over in the closet." She ducked back into the stall before he could regain his dignity. He walked to the closet, pulled out a hand broom, and started to brush himself again.

"Let me know when your back should be done," Grace called.

"Thanks," Matt said indignantly. He spent the next few minutes brushing with no intention of asking for help. Despite his silence, he heard the latch to the stall open and saw Grace step out. Her dress was wet.

She picked up the long broom. "Turn around," she commanded.

"I can do it."

"Turn around," she repeated.

"Fine."

She swatted at him fast and hard and he could see the dust in the air as it left him. "Were you able to sweep any off?" she asked as she brushed.

"There was no one around when I was done stacking hay."

"It's done," she said. "You're clean."

Matt turned around to look at her. Her face was gorgeous in the lamplight. He had to force himself to speak. "Are you working on Joshua?"

"Yes. I clean him in the evening and apply the apothecary ointment."

"Is it working?"

"No."

"Can I see him? Maybe there's something I can do."

"I don't think you can add any more experience than Uncle or Father."

"I thought that you weren't happy with what they suggested," he replied.

"I desire more," she said, "but no one knows horses better."

"I'm an apothecary, right?"

"But you know nothing of horses."

"Grace, let me see the horse."

She turned toward Joshua's stall and Matt followed. Joshua was a light brown stallion with a dark brown mane, similar in size to Shadow, but not nearly as muscular. Matt grabbed the lantern, walked into the stall and stepped close to the horse. Joshua had an infection on his back, which looked to have started under his saddle line. There were spots where the hair was falling out and scabs where the sores had started to bleed. The horse had a severe fungal infection that was being made worse by opportunistic bacterial infections.

"This is ringworm," Matt said aloud.

"I've looked," Grace said. "There are neither insects nor worms."

"Ringworm is a fungus."

"Mr. Miller," Grace said, shaking her head, "can you cure him or not?"

"I saw something along the road that might work for this," he explained. "We can give it a try in the next few days."

"We don't have a few days," she said. "Father could decide to put Joshua down any time."

"I'll try to come up with a cure."

*********

Matt wasn't hungry, so he went straight to the barn to wash. He filled two basins, brought them inside, made sure to close the barn door tightly, and then tried to bathe as best he could. He stood on a loose wooden board so as not to stand in the puddle forming on the dirt floor. He would have given anything for a hot shower.

He dressed and looked at his watch. He had plenty of time before he needed to be back for another text message, so he headed to the common. There were men sitting around the tables, as had been their routine. He grabbed a cup of cider and walked over to greet David and Will.

"How does it with you tonight, good fellow?" David said.

"It took forever to remove the hay dust," Matt replied.

"I still sneeze hay," Will said.

"Well done with the bales," David said. "You stayed until the job was done."

"Thanks," Matt said. "I like the look of a nice level stack."

"This is where we'll have you tomorrow," David said. "It seems the others know nothing of stacking hay."

"You get vermin in there too, if there is too much open space," Will added.

"Go to the stone field in the morning," David said, "then return to stack after lunch."

Matt nodded.

David continued. "We should have enough to fill the pavilion and then most of the barn."

"It's been a good season, then?" Matt said.

"Thus far," David answered.

"If it wasn't raining, would you work through the weekend?" Matt asked.

"No," Will replied. "Father encourages the men to go home and attend church."

"We want them out so the shacks can be cleaned and aired," David added.

"Were the shacks built as part of the original farm?" Matt asked.

"After, as slave quarters," David said.

"Must have been a lot of slaves," Matt observed.

"Sixteen bought and paid," David replied.

"Sixteen slaves!"

"A modest fortune," Will said. "Nathan took some horses and all the slaves as his share. He still has five of the original lot."

"We clean the shacks at week's end," David repeated. "You can help on Saturday if you wish. It would be the daily rate."

"I'll help if you need," Matt said, "but I'd rather go to Richmond. I'll check on my ring, talk to some merchants, and maybe buy some things." Matt planned to buy back his ring from Jacob Berkley and find a few antiques to take back with him when he was rescued.

"Your choice," David said. "I'm buying supplies in town on Saturday. You can ride in the wagon."

"What's this I heard about you not riding a horse?" Will asked.

"Maybe one of you will teach me to ride," Matt said. He smiled, thinking that it might be fun to have a few lessons before he was rescued.

Will motioned to a table. "I must sit down, my back's sore," he said. He and Matt sat down. David excused himself to talk to the men and eventually came back with two glasses of ale, which he set down for Will and Matt before leaving again.

"A nightcap," Matt said.

"A what?" Will asked.

"A nightcap. It's what they call a drink before you go to bed."

"A nightcap for me too, then," Will said as he reached up to rub his shoulder. "I'm baked."

"I brought something for you," Matt said. He reached into his pocket, pulled out two ibuprofen tablets and handed them to Will, who washed them down with ale.

"I hope they work as yesterday," Will said.

"The medicine prevents the swelling that causes the soreness in your muscles," Matt explained. "You'll heal more quickly."

"My muscles are hurt?"

"A little bit," Matt said. "Your muscles are sore because they are torn. Your body comes along and repairs the damage, and makes them bigger in the process. That's why men who lift heavy weights often get big muscles."

"How do you know all this?" Will asked.

"I learned it during my time at the university," Matt said. "I have a degree in medicine." Matt's degree was actually in pharmaceutical chemistry, but he wasn't sure whether the word *pharmaceutical* existed in 1762.

"I feel sorry for saying it, but it's time for me to sleep," Will declared. "I can feel the ibuprofen working already."

Matt walked alone back to the hay barn. When he got there, he turned his phone on and called for the dog. Scout came trotting up after a moment, zipped past Matt into the building, and jumped up on the bed. The dog was almost asleep when he lifted his head up at the sound of Matt's beeping phone. There was a new text.

"Working on your return," it read. "Any contact with the others?"

Matt typed, "No. Who am I looking for?"

"31 YO Brit engineer, 16 YO girl, 39 YO mom."

Matt typed, "No contact. Time for me to come home?"

"Working two months. No progress."

"It's only been a day."

"Only in your time. Two months have passed here. Wormhole is closing. Will text you again in exactly 24 hours."

Matt reached over to dim the lantern and then crawled into bed. He stretched his legs out, expecting the dog's regular growl, but the animal was quiet.

"Well, dog," Matt said, "you might be stuck with me for longer than I thought."

Matt lay there thinking of the implications of this latest message. He'd been agonizing on the trail over what he should say to his girlfriend when he returned after three weeks of being gone, but now it had been longer. They had both been too drunk that last night to have a rational discussion about their future. He doubted she'd respond well to his desire to change their lives anyway. Matt had used his hiking trip as an excuse to go home and get a good night's rest, but he knew that part of the reason was that he was not ready to deal with how she might react. Matt suspected he'd wasted eight months building a relationship with a beautiful socialite who had no intention of ever becoming a wife and mother.

Matt looked down at Scout and smiled like he was joking. "I could avoid all this hassle if I stayed in 1762." As soon as it came out of his mouth, a shiver went up his spine. It didn't sound that crazy.

# CHAPTER 15.

# GOLDTHREAD

---

The dog woke Matt before sunrise, scratching at the door. "I'm coming," Matt said. He put his feet onto the dirt floor, walked to slide the door open, then pulled his pants on and made his way to the privy. His mind was occupied both there and back with plans for his trip to Richmond. He'd visit the silversmith, retrieve his ring, and check out the apothecary shop. He was anxious to see the kinds of medicines that were stocked in 1762.

Once back at the barn, he got his backpack and wash-basins and took them to the well, set everything down and filled the basins with fresh water. He dipped his cup into the bucket and drank the entire contents. It was something his tae kwon do instructor had started him doing. The Korean man had made frequent suggestions regarding the importance of a proper diet, but this one had stuck. He could remember his sharp accent, like a staccato drill sergeant's: "DRINK FULL GLASS WATER – IN MORNING—GOOD FOR STOMACH."

His instructor, a seventh-degree black belt, possessed a multitude of wise Korean snippets and Matt found them coming back to him at the strangest of times. He was sorry that he hadn't had a chance to go back to visit the dojo in the two years he had been working in Philadelphia. Matt had completed his black belt as he was graduating from the university and then had to focus on work. There hadn't been time to join another tae kwon do school and he had only been able to practice a few times since graduation.

"Good morning, Mr. Miller."

Matt turned around to look at both Jonathan and Jeb. "What brings you here today, boys?" he asked, but he had a good idea of what it was.

"Jeb doesn't believe that you use soap in your mouth."

"Like I said, John," Matt explained, "it's toothpaste, not soap."

"It's Jonathan," Jonathan said. "Can you show him how you do it?"

"Okay, one last time," Matt said. "But if someone new shows up tomorrow, neither one of you will ever taste toothpaste again."

Jeb didn't look too excited.

"It does taste pretty good," Matt said, reaching for the tube. "Give me your finger." Jeb hesitated, but it was obvious he wasn't willing to let his little brother see him afraid to try anything. Matt placed a dab on his finger.

"Looks like soap," Jeb said.

"It's toothpaste!" Jonathan exclaimed. "It tastes good."

"He's right," Matt said. "It's not like soap at all."

Jeb raised his finger to his mouth slowly as Jonathan waited anxiously beside him. "I told you so," Jonathan said. Jeb swirled the flavor around in his mouth. "You're not supposed to swallow," the younger boy said. "Spit the bubbles on the ground."

Jeb spat the toothpaste out and said, "It tastes good."

"Let me show him how to use the brush!" Jonathan exclaimed. Seeing the boy's excitement, Matt knew there really was no way he could say no, so they spent the next few moments teaching Jeb how to brush. Matt stood by as both boys brushed their teeth.

Jeb said, "My mouth feels cold."

"Breathe in," Jonathan said. "It's refreshing. And the ladies love a man with fresh breath."

"I still don't think it's smart walking around saying that," Matt said, laughing.

"He should use it before he speaks to Sara Greene," Jonathan exclaimed. "He loves Sara Greene." Matt could see the older boy's face turning red.

"I don't love her," Jeb said.

"You do love Sara Greene. Use that brush before church." Jonathan didn't seem to be mocking his older brother. He sincerely wanted Jeb to get on well with Sara.

"Boys," Matt said, "time to go to breakfast."

"He loves Sara Greene. She's very beautiful."

"I get it," Matt said. "See you guys later." Matt pointed in the direction of the common. Jeb yanked at Jonathan's shoulder, pulling him in the direction of breakfast.

The boy turned towards Matt. "He does love Sara Greene. That brush might help." Jeb reached back and turned him around. Matt rinsed his toothbrush and inspected it. It had been new when he had started his trip, but already the bristles were starting to wear. He secured his pack in the barn and then hurried to the common.

Everyone was a bit grumpier this morning, with most complaining of being tired and sore. Matt ate by himself, having no desire to hold up his end of a conversation. The concept of leaving a good impression didn't seem as attractive now that he wasn't sure he'd be rescued any-time soon. The coffee tasted as good, and he couldn't eat enough bacon, but his enthusiasm had waned.

"Good morning, old fellow," Will said as he sat down beside him. "How'd you sleep?"

"The dog woke me up," Matt replied. "He shot out of the barn before sunrise."

"There were wolves west of us," Will explained.

"Does he chase them?" Matt asked.

"Stays behind the fence and barks," Will said, "which is probably wise against a pack of wolves."

Matt changed the subject. "Another busy day?"

"More of the same," Will said. "I grow weary of fighting hay."

"I'd complain if I wasn't getting all this fresh air," Matt said. It was comforting for Matt to see Will smile at his allusion to their earlier altercation with the Paynes.

"It would vex Levi much to fill the farm with guest workers," Will observed.

David interrupted their conversation. "Check your assignments," he called, setting a slate out on the table. Matt read that he'd been assigned to a new team headed by a man named Angus Stewart. David stopped Matt as he was backing away from the list. "Make sure you ride back after the first cutting. We want you stacking." Thinking for a moment, he added, "I'd eat with the ladies so you're ready to go. The wagons should begin arriving before noon. There's still one stack that's ready to fall."

"Which one is Angus Stewart?" Matt asked. David pointed to a stout man with enormous arms standing near a wagon. Matt went over to introduce himself.

The man shook his hand and replied in a thick Scottish accent, "Hallo, Mr. Miller, und I'm Angus Stewart. Arr ye ready to mow?"

"Let's get a move on," Matt said. "There's a prize waiting for us."

"Aye, that there is," Angus replied. Matt had to listen closely to understand. Stewart spoke like he had a mouth full of marbles. Angus waved to the other men standing nearby and repeated what Matt had said. "Come on, lads, let's get a move on, there's a prize waitin fer us." The five men climbed up on the wagon. Matt knew one of the men, Zachariah, from his previous work group.

"Another day in the fields," Zachariah said.

"It'll be a long day, too, if we want to get this done by Thursday," Matt replied.

"It'll be capital to return to town with a week's pay and an extra day before the Sabbath," Zachariah exclaimed. It was more enthusiasm than Matt had wanted.

"Aye, lad, that it will," Angus said as he turned his head from driving the two horses. "Of course, yer like as not to have your money spent by Friday evening, the way you carry on."

"Not true," one man in the back said. "He's an old man now. 'Twill last until Saturday evening." They laughed.

"What about you, Mr. Miller?" Angus said. "Where you intend to be spending yer money?"

"Nothing exciting," Matt said. "I need to pay for my trip back to Philadelphia."

"Philadelphia?" someone in the back asked.

"Born and raised," Matt replied.

"That explains the accent," said Zachariah.

Angus turned around. "Could you not make your way north already?"

Matt hadn't thought about answering this question, but it was suddenly obvious that he should have a cover that would explain his need to work while he waited for his rescue. A plausible explanation came to him almost immediately. "I'm saving," Matt explained. "I hope to be able to buy a horse and ride back to Philadelphia on my own."

"I have one to sell," someone said from the back.

"Old Brownie?" Zachariah replied.

"Yes sir," the man proclaimed. "He has one more trip left in him and he already told me he wanted to see Philadelphia before he passes." There was laughing.

"He'd never sell," someone said. "Everybody ponders where Caleb's silver goes. Providing for that horse in his old age, it is."

"He's a perfectly fine animal," Caleb declared.

"'Tis not for me to disagree!" Angus replied. He spoke to Matt again. "I'd say to buy one o' them Taylor horses, but they want a pretty penny."

"Man rides around Richmond on one of them animals," Caleb said, "and every beautiful lady will turn her head."

"Too bad a young man can't afford one," proclaimed Zachariah.

"Unless he's a plum," Caleb said.

"Wealthy fellows don't need horses to impress the ladies," Zachariah retorted. "They got their relations' money."

"'Tis true, lad," Angus said. "Fine horses are wasted on the old and the wealthy, they are."

"Hear! Hear!" the men cheered in the wagon. Matt smiled, thinking that young men in his own time could say the same thing about Porsches.

The conversation had ended by the time they pulled up to the field. Matt dropped over the side of the wagon, walked to the back, chose a scythe and a sharpening stone, and then tried to listen as Angus barked out unintelligible instructions. He followed the pack and situated himself towards the middle of the group. His third day in the field was much easier; sometimes he found himself ahead of both Caleb and Zachariah and would wait with Angus for them to catch up. Despite the mismatched productivity, the team developed a rhythm that regularly corrected itself, each man having enough time to pause, sharpen his blade, and take a moment to rest.

It took about two hours to cut the field and spread the hay to dry. By the time they finished, David had arrived and was waiting in a wagon. Angus spoke to David briefly and then stepped to where the men were cleaning their tools and stacking them in the cart. "Philadelphia is going back to stack," Angus announced.

"Better him than me," Caleb said. "Hard to breathe in there."

"Too much dust," Zachariah said. "I'd rather be out in the field."

"You men enjoy the outdoors," Matt said. "I'll be sweating and coughing." He tried his best to act like he was making a sacrifice, rather than doing something he actually enjoyed.

"Good work this morning, Mr. Miller," Zachariah said as he was leaving. "If I didn't know better I'd say you had a few harvests under your belt." The comments made Matt proud. He liked the feeling of teamwork and was glad to contribute.

"Come on," David called. "You can puff like dandy Frenchmen another time." Matt climbed aboard the hay wagon along with David.

"Would you mind driving to the north hayfield?" Matt asked.

"We have all that hay," David replied.

**Common Names:** Goldthread, Yellow Root, Mouth Root, and Canker Root
**Scientific Name:** Coptis groenlandica
**Family:** Buttercup (Ranunculaceae)
**Flower Color:** White

"It's only about halfway. There's a place along the road where a medicinal plant is growing."

"Tomorrow would be better when it's raining. You'll have all day to cut plants."

"A plant is growing along that road that may cure Joshua," Matt said.

"Don't fill that girl's head with false hope," David warned. "I've never seen skin corruption this severe. I pray it doesn't spread before my brother puts him down."

"I have seen the same disease on a man and it was cured by this plant," Matt said.

Matt was lying. He hadn't seen goldthread cure anything, but he had read a research report about its antifungal properties. He remembered it because it looked like the honeysuckle flowers he and his friends used to pick in abandoned lots to suck on their stems. They probably tasted every white flower in the neighborhood trying to feed their honeysuckle addiction. When he saw goldthread in that report, it immediately registered, and it had flashed into his mind again when he saw it on the way back from the hayfield.

"Grace doesn't have to know," Matt said. "Give me a chance to try."

"The road is up on the left," David said, resigned. "Truth be told, none of us will be able to bear that girl if she loses that horse."

They drove for longer than Matt expected and the terrain looked very different heading in the opposite direction, so for a moment he worried that they might not find the plants, but they soon were traveling through a sea of white flowers. Matt pointed them out and David stopped the wagon. "Go quickly," David said.

Matt hopped out and grabbed one of the smaller scythes and pulled a burlap sack from underneath some tools. He walked over into the flowers, which were mixed with a plush carpet of green moss. The flowers resembled those he remembered from the report, but he had expected to see the thin gold stems that contained the medicine. There was nothing gold about this plant. He brushed around in the plants, but didn't see what he wanted. *This isn't it!* He kicked at the plants in disgust with the toe of his boot, pulling a few of the plants out of the ground. To his surprise, a bright golden yellow root popped from the ground. *The roots are gold!*

Matt set the scythe down, got on his knees, and began pulling plants from the ground. When his sack was full, he stood, trotted over to the wagon, and tossed everything into the back. "I got what I wanted," he said as he hopped into the seat next to David.

As they rode, Matt tried to formulate a plan. He'd need a way to grind up the goldthread root into an oil or cream and extract the active ingredients. There would have to be enough to spread over the horse's back for the next few weeks. David stopped the wagon in front of the horse barn and Matt jumped off. He grabbed the burlap sack from the back of the wagon, put it over his back and started toward the barn. "Make sure you get a meal in," David called after him. "They arrive with the hay soon and you still need to restack those back bales."

Matt found an upper shelf in the barn and stuffed the sack securely onto the ledge. Most of the horses poked their heads out of their stalls, looking curiously at him, except for Joshua, who was conspicuously missing. Matt walked quickly down to his stall, scared of what he might discover, and was relieved when he saw Grace standing next to the horse holding a bloody towel.

"There's naught I can do," she said quietly. "His hair falls out in patches."

"When's the last time he was in the sun?" Matt asked.

"Not since his infection got worse," she replied. "Father asked that he be kept from the other horses. He spends most of his time under the shade trees anyway."

"He needs to be outside in the sun," Matt said. Every scientist from his own time knew that exposure to UV light would kill bacteria and fungi. Matt motioned to Grace to follow. "It'll help to heal this," he said. "Come on."

"Where would you have me go?" she asked, puzzled.

"I need your help," he replied, "to build a small corral out in the sun." Shrugging, she stepped through the open door of the stall and closed it behind her as she followed him into the large corral. She watched him for a while in silence as he disassembled fence sections and set wood on the ground to mark the perimeter of a smaller corral about three times the size of Joshua's indoor stall. They worked together to join the fencing.

The corral was mostly finished before Matt noticed how physically close he was to Grace. Despite her proximity, he was able to maintain his focus until the very last fence rail. They both had to hold the back end of the rail outside the completed circle to push it into the space that remained. It got wedged between two uprights and they had to yank it backwards. She fell into him when she lost her grip and he caught her in one arm while still holding the fence rail with the other. She glanced over her shoulder while in his arm, held his gaze for a moment, righted herself, and grabbed the rail again. This time they were able to push it into place.

"I'd put a blanket on Joshua so he doesn't touch anything on the way out," Matt said. "Let's bring him into the sunlight." Matt waited for her to lead Joshua from the barn into the corral. With the blanket covering his back, you would not have known anything was wrong with the horse. He bucked slightly, though, as it was removed and tore at the scabs. Once free, the horse pranced around the corral, glad to end his quarantine.

"The sun should help," Matt said. "I'll mix medicine when I'm done stacking hay. Don't use any more of that cream you have and don't get him wet. I want him as dry as possible."

# CHAPTER 16.

## HORSE SENSE

---

Matt could only think about curing the horse as he stacked hay. He hoped the goldthread could be applied before the sun went down, but he needed to extract the antifungal chemicals from the roots and find something like petroleum jelly to make a suitable ointment.

By the time dinner came, Matt had his plan, so when the last hay bale was in place, he rushed to the common for his meal. He filled his plate, grabbed a fork, and carried his food to the farmhouse and into the kitchen. Grace was there with her mother at a kettle over the fireplace. Flames were burning in the hearth and licking at the cast iron pot. The women looked like witches working on a potion, and this made him smile.

"Grace," he called out.

She had a large spoon in her hand and looked comical standing there in her apron. His smile widened. She wasn't expecting a visitor, especially a smiling one.

"Mr. Miller," she said, "I'm busy."

"Can we talk for a moment? It's about Joshua."

Mary spoke as she looked back and forth between them. "I can do this myself, dear. Go speak with Mr. Miller about your horse."

Grace set her spoon down and walked outside with him. "What now?"

"I want some things to make the medicine for Joshua's back."

"What things?"

"What kind of oil do you have on the farm?"

"Sometimes there's olive oil we buy in Richmond, but it has not been available," Grace explained. "It's costly, so if we did have any, Mother wouldn't let you take it."

"How about cooking oil?" he asked. "Corn oil?"

"Oil from corn?" she said, puzzled. "What kind of corn has oil in it?" She went silent, waiting for his answer. She looked beautiful.

*Beauty!* "What about women's beauty cream?"

"Beauty cream?"

"You know, cosmetics."

"We don't have any beauty cream, or *cosmet...*or whatever you called it," she said.

Matt could see the impatience growing on her face. "Do you use anything for dry hands?"

"There's sheep butter in the barn," she said. "We use that in the winter."

"Sheep butter?"

"Yes," she replied. "I don't favor the smell."

"From sheep's milk?"

Grace laughed heartily. Matt looked at her, bewildered. "You *are* a city boy," she said.

"You said sheep butter," he replied indignantly.

"They press it from the wool."

"Can you show me this sheep butter?"

"Let me tell Mother." She pulled off her apron and hung it on a peg beside the door, and went back into the hearth room. She was gone for a moment, then walked past him in full stride, motioning for him to follow. "Did the stacking go well?" she said when he finally caught up to her.

"We filled it," Matt answered. "The rest will have to go into the hay barn."

"Stack it tight," she said, "or you and Scout will want for other lodgings."

"We'll figure it out," Matt said. "Me and the dog have become friends over the last couple of days."

"Sometimes I wonder about that animal," she replied, laughing. "He'll befriend any passing stranger."

"Not just anyone," Matt replied, willing to meet the jest. "We've worked out a deal."

"What's the deal?"

"I make sure we have room to sleep and he lets me pet him without tearing my hand off."

"I have bandages in the house when you need them," she replied.

Once they entered the barn, Grace took him to a crudely fashioned tin container that was about the size of a gallon milk jug. She worked the lid off with her fingers and then stepped back. The container was about half full of milky, amber-colored grease with an animal odor, that wasn't as bad as she had described. Matt smashed some between his fingers.

"This is exactly like hand cream," he said.

"I told you we apply it to our hands in the winter."

"You did…" He trailed off. He was thinking aloud. "Can I have this?"

"For what?"

"I have a medicinal plant to grind with this. We'll apply it to the horse's back to cure him."

"Why can't you say you're making an ointment?"

Matt reached up and grabbed the bag of goldthread he had collected earlier in the day and pulled out a few of the plants to show her.

"You better get these ground before the medicine goes bad," she said.

"You may be right," Matt said, looking down at the wilted leaves. "I'll need a knife, a bowl, and something to grind in. Maybe a heavy spoon for grinding that fits the bowl." *It's not likely they'll have a mortar and pestle lying around a horse farm!*

Mary was still stirring her cauldron when they returned to the kitchen. Grace searched through a number of cabinets as Matt watched and then set a knife, a wooden spoon, and a metal bowl on the table.

"I'll get started right away," Matt said. He grabbed the stuff and was ready to walk out the door.

"Wait," she said. "I'm coming with you." She looked at the items in his hands and said, "There's something else."

Matt was impatient. "What's that?" He had already started working out his plan and was only half paying attention. She could add little more than delay.

"A mortar and pestle," she said. He stared at her in disbelief and set the things in his arms down on the table as she went to a cabinet, pulled out a large mortar and pestle, and placed them in front of him.

It felt like Christmas Day. "You brilliant woman!" Matt exclaimed. He reached out and hugged her. He had his arms around her when it occurred to him that a hug might not be an appropriate thing for a man to do to an unmarried woman in front of her mother in 1762. When he let her go, Grace stood there, arms at her side with her signature smirk on her face. She glanced over at her mother.

Matt looked first at Mary and then back to Grace to explain himself. "I thought that there was no possible way that there would be a mortar and pestle on the farm," he said. He knew he was lying to both them and himself. The opportunity had presented itself for him to wrap his arms around this beautiful creature and he had taken it.

Mary let it pass. "God will provide, Mr. Miller," she said simply.

Matt scooped up all the supplies in his arms. "I have to get to work."

"*We* have to get to work," Grace corrected.

# CHAPTER 17.

## IMPROPRIETIES

---

There was plenty of light streaming through the windows; Matt and Grace would be able to work there for at least two hours before it got too dark. Matt took the bag holding the goldthread and emptied it onto the bench, grabbed a handful and lined up the plants so the roots were together, then used a knife. "Cuts easily," he said. "We want the gold roots. It doesn't matter if you get a few leaves in there." He pushed the knife over to her on the workbench, grabbed the cut roots and stepped to the mortar and pestle. He put the roots into the bowl and began to grind.

"Good day!" said Jeb as he walked into the barn followed by Jonathan. "We came to help."

*I don't have time for these jokers.* "There's not much to do," Matt said. "It's only getting these plants cut and mixed with the sheep butter. It's a two-man"—he looked at Grace—"err, two-person job. It'd be a great time to get chores done instead of hanging out here."

"What does *hanging out* mean?" Jonathan asked.

"Waiting around without doing any work," Matt said.

"We could do something," Jeb said.

"There's nothing right now," Matt replied. "I'd get your chores done."

"We aren't allowed to leave," Jonathan confessed.

"Jonathan!" Jeb exclaimed.

"Why aren't you allowed to leave?" Matt asked.

"Mother said we must stay to prevent improp, um…improp…prieties."

"Improprieties?" Matt asked. "Uncle David knows we're treating the horse."

"It's the improp…prieties with our sister," Jonathan corrected.

Grace stopped cutting the goldthread and announced clearly to everyone, "There's no chance of improprieties. Mr. Miller isn't one of my suitors, and he's certainly not a man that I'd ever consider marrying." It was as calm and lucid as anything he had heard her say since they had first met.

"And your sister isn't a woman I'd ever consider marrying either," Matt replied.

"Mother said you would say that," Jeb replied. "We're still not allowed to go."

"Fine," Matt said. He handed the mortar to Jeb. "Start grinding." Matt stepped to where Grace was cutting. She looked up at him with her smirk, having enjoyed her declaration. He grabbed the cut stems, took them to Jeb, and dropped them in the bowl for grinding. "You want to smash the stems up completely."

"What can I do?" Jonathan asked.

"What I was doing," Matt said. "When Jeb is ready, bring over more cut roots for grinding."

Matt had delegated the whole task, so there was nothing left for him to do but supervise. It took about thirty minutes to grind the plants with everyone working together. Jeb had been transferring the ground-up roots and sheep butter into a separate bowl, which was now filled with a thick yellow paste.

"We're finished," Matt said. "Let's get Joshua."

They walked out to Joshua's corral, where the horse was trying to eat plants that had grown just out of reach around the fence. "It would be nice to have a sheet to cover his back once we put the ointment on," Matt said. "Something old, but clean."

"Mother has torn linens we can use," Grace said. "I'll be back." She walked to the farmhouse, leaving Matt alone with the boys. The whole impropriety discussion was still weighing heavy on his mind.

"Who does your family want Grace to marry?" he asked.

"Father likes Robert Marsh," Jeb said. "He comes from a wealthy family. They have a tobacco business with plenty of slaves."

"What does Grace think?" Matt asked.

"She doesn't think he's very handsome," Jeb said.

"She said he's funny-looking," Jonathan offered.

"Women," Matt said coldly, shaking his head.

"He does look like a gopher at times," said Jeb. "He's a very nice man, though, and always brings candy when he comes to call on Grace." Jeb continued without any prompting. "Mother likes Daniel Sutherland, but Father thinks he doesn't work hard enough."

"Father likes men who come from good families and work hard," Jonathan said.

"Probably like most fathers," Matt replied.

"Do you come from a good family?" Jonathan asked Matt.

"My parents are divorced, and they aren't wealthy," Matt replied. "Probably no."

"What does divorced mean?" Jonathan asked.

"It means that a mother and father aren't married anymore," Matt replied.

"Your mother and father aren't married?" asked Jonathan. The boy looked like he had seen a ghost.

"I guess you don't have much divorce down here," Matt said.

"I don't know anyone whose parents aren't married," Jeb said. "Some have a mother or father who has died, and they often want help from the church."

"If your parents are divorced, do they still live in the same house?" Jonathan asked.

"One parent moves away," Matt said.

"Who cooks dinner?" asked Jonathan.

"Usually the parent that remains," Matt explained. "Or they buy food and bring it home."

"At least there's something for the children to eat," Jeb said. "If you like the parent who moves away, can you live with them?"

"Some do," Matt said.

"I'd not move out of our home," Jonathan said. "I'd miss everyone."

"Children usually spend time with each parent," Matt explained. "They go to one home, then the other."

"I'd live with Mother," Jonathan said. "Father is too strict."

"Your parents would never get divorced," Matt replied. He was sorry that he'd brought any of this up.

"Would it not be better if the children stayed in their home and the parents came to visit?" Jeb asked.

*That's a thought.* "I'm done talking about divorce," Matt said. "It'll never happen to you anyway, so don't worry."

"Will you ever do this divorce, Mr. Miller?" Jeb asked.

"Never," Matt said with all conviction. He had made this promise to himself long ago, and it was something he intended to talk to Kylie about when he returned to his own time.

"Capital," Jeb declared.

Grace returned and looked at them suspiciously. "Why is everyone so quiet?"

"Mr. Miller told us about divorce," Jonathan said. Matt was tempted to shush him, but he reconsidered, thinking that it would only make it worse.

"Divorce?" she asked, looking at Matt. "It's inappropriate to speak of divorce in front of young men." She seemed sincerely concerned.

"I didn't want to," Matt said. "They had a lot of questions."

"Mr. Miller said he'd never get a divorce even though they do it sometimes in his family," Jonathan said.

"Jonathan, I told you that Mr. Miller isn't a man I'd be interested in marrying," Grace said. Her face was turning pink, either with anger or embarrassment. "Therefore, I have no trepidation concerning Mr. Miller's attitudes towards marriage—or divorce."

"I was just saying it if you were interested," Jonathan said, now looking at Matt. "He works hard and doesn't look at all like a gopher."

"Thanks, Jonathan," Matt said. "Speaking about work, there's a horse to cure."

"I'm all for that," Jeb said. All the back and forth seemed to be wearing on the teen.

"I have the linens," Grace offered, as relieved as anyone to move from the present topic.

Matt led the group out to Joshua, who had now been out in the sun for a large part of the day. This wasn't enough time to see an effect on the infection, but Matt was happy to observe that the sun had dried some of the sores. Matt grabbed the horse's halter and motioned to Jeb. "Keep him still while we spread the ointment on his back."

"Jonathan," Matt called. "Can you hold the bowl while Grace and I spread it on his back?"

"Sure," Jonathan replied. He walked over holding the bowl in both hands, propping it against his belt.

They began covering the sores with the goldthread mixture. After each had finished their respective side, they reached together to apply ointment to the top of the horse's back. Matt's hands regularly brushed up against Grace's. To him, it was like an electric shock every time their hands would touch, but Grace didn't seem to notice. It occurred so often that Matt wondered if it might not be an accident.

"I'm done," Grace finally said. She looked at Matt with her ice-blue eyes.

"Me too," Matt said. He had finished much earlier but became too caught up in the motion of her hands to stop spreading. *I feel like a lovesick teenager.*

"I pray this works," Grace said, still looking directly into his eyes.

Matt grabbed a horse-sized piece of linen, wiped his hands off on it, and put it over Joshua's back. He took the twine and tied the underside of the sheet around the horse's belly as if he was securing a saddle. "That should do it," he said, stepping back to view his handiwork. He looked down at his fingers. "I'd feel better if we could go wash our hands with soap and water." He looked at Grace. "Any soap in the barn?"

She nodded and replied, "Join me at the well."

Matt headed to the well after grabbing the water basins and was able to fill them with water by the time she arrived. She set a large block of soap down next to him. "I've done this before as part of my job," Matt said. "Watch."

"I know how to wash my hands," she said.

"Humor me so you don't get some strange hand infection."

"I'll do my best," she said, rolling her eyes.

Matt ignored her. He took the soap and used the first basin to wash his hands once and then again like a surgeon, and then instructed Grace to pour water over his hands from the second basin to rinse off the suds. He dried his hands on a towel he had placed next to the basins. "Now it's your turn," he said, handing her the soap. She reached into the basin, wet her hands and started washing. She finished up quickly and motioned to Matt for the rinse water.

"That's not good enough," Matt proclaimed.

"My skin will crack if I wash my hands too roughly," she replied.

Matt picked up the soap. "Give me your hands," he said. He had visions of her skin becoming infected and her family fretting over the damage he had done to their daughter. She put her hands over the basin and Matt reached out to hold them in his while he used the bar to coat them with a fresh lather. He dropped the soap so he could use his fingers to make sure to scrub her hands completely. "There's ointment between your fingers," he said as he worked the soap around her fingertips and under her nails. "This needs to be off your hands, even under your nails—"

He was so caught up in the task that he was nearly finished before he looked up at Grace's face and realized that he had both her hands in his and she was staring directly into his eyes. Matt stopped midsentence and then it seemed like everything stopped; no speaking, no scrubbing, and no breathing. It was as if time had shut down; his world became completely quiet as he held her hands, motionless.

Matt had no idea how long it took Grace to look down and slide her hands slowly away. "They're done now, Mr. Miller," she said quietly. They finished washing in silence. When her hands were dry, she stood up, turned, and walked away without a word.

**********

Scout was waiting for Matt when he returned to the hay barn. "What's up, Cujo?"

Scout looked at him and tilted his head. The dog followed him into the barn while Matt checked his watch and glanced at the charge left on his phone. He pulled the kinetic charger out of his bag, wound it and connected the wire.

"I have time for an ale," Matt said to Scout. "Let's see what's going on." He motioned to the dog to follow as he walked out the door. Scout followed him all the way to the common, then peeled off and took his regular spot on the porch of David and Faith's house.

"I was sure we lost you to the bed tonight," David declared when he saw Matt.

"Almost," Matt replied. "I thought I should check to see what's on for tomorrow."

"You'll stack again," he said. "Return midmorning. We've another day with the hay."

Besides David, there were no other Taylors in sight. Matt suspected that they had finished their prayers while he was working with Joshua. After finishing his drink, he walked alone back to his hay barn. He glanced at his watch and saw that he still had time, thought twice, grabbed the basins and headed out to the well. He was starting to smell and now seemed like as good a time as any to wash. When he returned with fresh water, the dog had already let himself in and was lying on the foot of his bed.

"Hello, dog," he said.

"Awwarrrr." The dog let out a big yawn, opening his mouth wide.

Matt laughed. "That was about the most ridiculous sound I have ever heard from an animal." Scout laid his head on the bed and watched while Matt rambled on about the day's events as he washed, toweled himself off, and then dressed. He'd finished dressing by the time his phone notified him of his text message.

"Are you there?" it read.

Matt typed, "Any luck bringing me home?"

"We can get you back the same way you came."

"I was unconscious for two days and have bad headaches."

"Expect similar effects on return."

Matt typed, "Any way to test?"

"Don't think. One shot. Contact you in 24 hours."

"Test it first?"

"Sorry, working for six months. Best we can do."

Matt typed, "Not stepping into a wormhole without knowing it's safe." He didn't get a reply. He looked over at Scout. "Best they can do?"

# CHAPTER 18.

# MAD MONEY

---

Jonathan arrived in the morning as Matt was washing his face. "Good day, Mr. Miller," he said. "Did you sleep well?"

"Very well," Matt said. "I went to bed early last night."

"Is it scary in the barn at night?" the boy asked.

"Nah. Scout's my protection."

"Scout likes you," Jonathan said. "He doesn't go near any of the other men. It's only our family and you."

"I'm still afraid to pet him," Matt replied.

"He'd not bite. He likes you more than he likes Father. Father is very strict." He said "strict" with a particular emphasis.

"You have an incredible father," Matt said.

"He doesn't smile anymore."

"His daughter died," Matt said. "Give him a break."

"He won't let Grace ride anymore," Jonathan said. "She used to take me for rides with Kathryn."

"Grace rode a lot?"

"Until Kathryn got thrown," Jonathan explained. "Now she's not allowed because she'll fall. She's not allowed to wear breeches, either. I'm not supposed to talk about that."

"Can't talk about breeches?"

"Kathryn wore them, too," the boy said. "I don't see the bother. They didn't look like fellows, not one bit."

"Your secret's safe with me," Matt replied. "Where I come from, ladies wear breeches all the time. I don't see what the bother is either." He could imagine Grace walking around in men's breeches, causing a scandal in 1762. It wasn't too hard to imagine that the pants didn't make her look like a *fellow*. The vision of Grace in his mind caused a stirring; he longed to see her again.

"I suppose you want to brush your teeth," Matt said, changing the subject.

"If it's fine."

"It's fine," Matt replied. "Where's Jeb?"

"I asked him to come, but he said he didn't want to this morning. He's waiting for Sunday."

"Sunday?" Matt asked.

"When he sees Sara Greene at church," Jonathan said.

"I'm glad to help."

"Do you want help with Grace?"

Matt looked at Jonathan out of the corner of his eye. "I'd rather talk about something else."

The boy took the cue, thought for a moment, and asked, "Will the rain wash the medicine from Joshua's back?"

"Decent question," Matt replied. "The horses usually spend their time in the rain, right?"

"Unless it's overlong and the pasture is muddy, or if there's lightning," Jonathan said. "Father brings them in if he thinks they'll get muddy."

"Let's walk over there before breakfast and get that linen off his back," Matt said. "I want as much sun as possible on it before it rains." Then he added, "If it rains."

"Uncle says it will rain after dark tonight," the boy said. "He's never wrong about rain, but we like to joke with him when he picks the wrong day." The boy looked up at the sky. "The sky is still bright, so the light can heal Joshua."

"Let's get over there," Matt said as he was putting his pack together.

"I want to brush my teeth," the boy said.

"Oops," Matt said. "I forgot, but not on purpose." Matt handed Jonathan the toothbrush and waited patiently for him to complete the task. Once the boy was done, Matt gathered up his shaving kit, put it into his pack, and walked back into the barn to tuck it into one of the bins. They walked together to Joshua's corral. When they rounded the corner, they saw that Grace was already there.

"I'm taking the linen off so the sun can get in," she said.

"We came to do that," Matt said. "Make sure you wash your hands."

"With soap," she said. "Unless you endeavor to instruct me again."

"You've got it," Matt said. He made a conscious effort not to linger, and walked away.

The boy talked the entire distance to the common but split off once he saw breakfast. "Good day, Mr. Miller," he said as he trotted away.

"Looks like you have a fellow," Will said, watching the boy leave.

"He comes by to give me advice," Matt said. "Life on the farm and such."

"He asks how you did in the fields," Will said. "He wants you to do well."

"Not sure why," Matt said, perplexed.

"He found you," Will replied.

"Found me?"

"Under the bridge," Will explained. "Jonathan saw you first. If you cure that horse, he'll be taking full credit."

"I'll try not to let him down," Matt replied. This new information explained a lot.

"Is your medicine working?" Will asked.

"Grace is pulling the linen off now," Matt replied. "I'd be surprised if we see anything this early."

"We can pray," Will said. He turned to the gathering group of men. "I'll probably see you this afternoon. Tobacco is mostly done."

"What was the final assessment of the crop?" Matt asked. "Was it worth the time?"

"It's in God's hands," Will said. "We'll not learn until we sell."

"When's that?"

"'Twill take some time to dry. It can't transpire quickly enough for Father, though. He already has his eye on a horse at the Browne farm."

"Expensive horse?"

"Father and Uncle have schemed for months to justify the cost."

"The tobacco crop, it's your mad money," Matt proclaimed.

"Mad money?"

"In Philadelphia, we sometimes call money that you would not ordinarily have 'mad money,'" Matt explained. "It's the extra money that you don't require for your regular expenses." He looked around, motioned with his arms at the surrounding farm, and then went on. "You can use it on things you wouldn't normally buy. You know, to reward yourself for extra work or being smart."

"Like a horse that most consider too costly," Will offered, smiling.

"Exactly," Matt exclaimed. "Mad money!"

"Mad money," Will repeated as he nodded in agreement.

# CHAPTER 19.

# CORN

---

Wagons came and went the whole morning and Matt lost track of how many they filled with corn. The men on his crew were anxious to finish and get back to town for their three-day weekend, so most worked in relative silence. Compressing five days into four had sounded like a good idea, but the tedious work had taken its toll and they were all growing weary.

Matt spent his fourth day in the field alongside a young man named Francis McKean. Matt had introduced himself to Francis in the morning and was hoping to make another friend, but found within the first hour that Francis wasn't interested in conversation. After getting about ten separate yes and no answers, Matt gave up and resigned himself to pulling corn from stalks and whistling all the Beatles songs he could remember.

David, Mary, and Jeb arrived with the meal wagon around noon and they stopped for lunch. It was darker today now that clouds filled the sky. David said, "I believe you men will be going home by early afternoon. Let's shorten the break and get as much corn in wagons as we are able before the storm." Matt noted that the forecast had turned from rain to storm. He gazed up at the clouds and inhaled the smell that was now heavy in the air. David looked at Matt and said, "I'll take you back. The pavilion's full. The rest goes into the barn. Anything that gets wet stays outside." He looked to the clouds to emphasize his point. He paused for a moment and smiled. "Grace said to tell you she thinks that cure of yours is working."

When they arrived, hay was already in front of the barn, so Matt was unable to go over and see the horse. As he was stacking, he saw Scout meander up to the barn. The dog sat watching as Matt moved the hay from one pile to another. "Have you checked Joshua today?" Matt asked the dog. At the sound of the horse's name, Scout turned his head to the horse barn. "I heard he might be looking better."

"I don't know if the dog has seen him, but I have," Grace said, walking around the corner. Matt was mildly embarrassed that she overheard him having a conversation with the dog, but he still gave her his best "I'm proud I talk with animals" look.

"Have the sores started to heal?" Matt asked.

"The weeping has stopped, and the swelling is down," she said. "Something is happening." She then added, "He may be getting better on his own."

Matt knew she was trying to antagonize him, but he wasn't biting. "Well," he said, "if that's true, you can probably throw the rest of that medicine away." He smiled calmly at her and reached down to pick up another hay bale.

"There's still the slightest chance he may be getting better from our ointment," she replied. She raised her hand and put her beautiful thumb and forefinger together to illustrate a very tiny open space. Matt could only shake his head and laugh. "Slightest chance," she repeated, smiling in such a way that he wasn't sure what she believed.

"Did you come to help me stack?" Matt asked.

"Men's work," she replied. "I'll leave you and your fellow here to finish." Scout's eyes followed her all the way to the house.

"Can you believe her?" Matt said to the dog.

"Believe what?" said a voice from the other side of the barn. David rounded the corner and looked at Matt expectedly. "What was that?"

Matt had to explain himself a second time. "I was talking to the dog."

David looked in wonder at Scout. "It's odd to see that animal linger about. Bating the time he spends with Thunder, we usually don't see him."

"Thunder's that large horse he plays with in the pasture?"

"They grew up together," David replied. "They carry on endlessly if left to their own wits. It's like watching children play. Ofttimes that dog will stand outside Thunder's stall and torment him to no end."

"Can't you keep him away?"

"Wouldn't dream of it," David replied. "They both do their share. Thunder is still last in when the dog is rounding them up—wanders away on purpose." Scout was turning his head back and forth as David and Matt spoke, like he was part of the conversation. David looked at the dog and said, "You know *Thunder*." The dog's ears perked, and he stood and then trotted away.

"Sorry about that," David said. "You've lost your fellow."

"He'll be back tonight to bother me," Matt replied.

"He keeps you up barking?"

"He sleeps with me in the barn."

"All very strange," replied David. He had a hint of suspicion in his voice.

"Scout will be glad when I return to Philadelphia," Matt said. "He'll get his barn back."

"When do you plan to travel?" David asked. "Not before we are done in the fields, I hope."

"Soon. I'm still trying to figure out how I got under that bridge."

"A man doesn't find himself lying in the road with no picture of where he was before," David quipped.

"The only memory I have is leaving work to explore the countryside," Matt explained. "The next thing I remember is waking up in your barn."

"How long have you been gone from your home?"

"No idea," Matt said. "My story sounds suspicious, I know."

David nodded affirmatively. Matt had suspected as much. "I appreciate that you've trusted me so far, all of you," Matt said. "I'd be in pretty dire straits if you hadn't helped."

"I don't want to give the notion that I mistrust you, it's just . . ."

"Finish, David," Matt said. "It's important for me to know where I stand."

"The things you say and do don't fit," David replied.

"Like what?" Matt asked. He wouldn't have been surprised if David had a laundry list of things Matt had said and done that didn't make sense. *It comes with the territory when you get dropped into another century against your will.*

"I don't desire to interrogate you," David replied. "You've got a disarming manner, but I can't help thinking that we don't have your full story."

"I'm trying to get back home," Matt said. "You can trust that much."

David's manner changed, almost like he made some conscious decision not to pursue the matter further. "There's still a lot of hay," he said. "Best get it under shelter."

"I'll get right to it," Matt said, smiling. "I'm proud of my haystacks."

"Get some victuals in your stomach before they're put away," David reminded him. "We don't want to wind up peeling you off the ground again and having to reteach you farming."

**********

Two full wagons of hay had already been unloaded for him to stack by the time Matt had finished his meal. He set his plate down, still chewing, and started at his task again. When the pile was gone, he finally walked to the corral to check out Joshua's back. He didn't expect much as he walked up to the animal, but was happily surprised.

# CHAPTER 20.

## RIDING LESSONS

By dinnertime, the farm had grown eerily quiet except for the sound of the pattering rain. The last wagon headed for town was leaving the front of the corral and the rain was falling steadily. Matt stood in the entrance to the hay barn looking at his handiwork. He had organized the bales into a stack that occupied less than one third of the building. There was plenty of space left for equipment, tools, supplies, and even more important, for him and the dog to sleep. Looking at the wide-open space, Matt had to consider that all the talk of filling the barn completely with hay was a complete exaggeration. They would've needed another couple of hayfields. He breathed in deeply, enjoying the smell of the freshly mown hay.

Matt turned around to look over at the corral to Joshua, who was standing under a lean-to that David built to keep him dry. The horse stooped down regularly to grab a mouthful of hay. He seemed to enjoy being pampered under his shelter, and the linen sheet on his back made him look like a thoroughbred racehorse after winning the Triple Crown. Matt could see activity in the nearby horse barn with various Taylors coming and going. Now that his work was done and Matt was left alone with his thoughts, he was having trouble coping with the solitude. In his own time, he'd be checking his e-mail, reading the news online, or updating his status. There was nothing for him to do now and it gave him a strangely uneasy feeling, especially after moving at the speed of light for six straight days.

The last text from Oak Ridge said six months had passed since he vanished from his own time. Matt's mind drifted to wondering who had sincerely missed him after he'd disappeared. The only person he could be sure of was his father. His old man would often pick Matt up at work and they would eat lunch together in the city. He thought about their last meal, when they rode in his dad's new Prius taxicab, talking as they waited in Philadelphia traffic.

"If I drive it right," Matt's father said, excited, "I can get almost fifty miles to the gallon."

"It's still a Prius," Matt replied.

"Wouldn't impress that girl of yours," his father said.

"Drop it, Dad," Matt replied. "I know you don't like Kylie."

"I like her fine," the older man said. "It's just...she reminds me too much of your mother. Beautiful and self-absorbed."

"Like father, like son," Matt quipped. Matt hadn't felt like fighting fair.

"You'd think you'd learn from your old man's mistakes," his father said. "She want a family at all?"

"We never really talked about it," Matt lied.

"After seven months?"

Kylie *was* beautiful and self-absorbed. She had expressed her disdain for children on more than one occasion. Matt had to admit that much of his attraction was based solely on the fact that she was beautiful and the life of the party. Matt liked the way that people, men and women both, turned to watch when they entered the room. As hard as he tried, though, he couldn't imagine Kylie as a caring mother of his children. If she did have children one day, Matt suspected she would always have a nanny in tow. He wondered what Kylie might be doing now that he was gone. His conclusion was sad; she'd probably be going to parties and having fun, which is what she did best. *Kylie missed me for maybe...a month?*

He thought about his job at the pharmaceutical company and decided that his disappearance had briefly caused an inconvenience, but they'd have either quickly filled his position or had some underpaid schlep in India do it for one quarter of the price. *They missed me for as long as it took to ship my job to Bangalore.*

The fact that he couldn't think of one person besides his father that gave a damn whether he was gone or not was troubling. *If I get back, things are going to have to change!* He stood there and watched the rain fall, then decided to walk to the horse barn to see if he could help. He was getting tired of trying to deal with the ambiguity of his future when he returned to his own time. When he arrived at the barn, Grace was leaving to go back to the house.

"Hi, Grace," he said.

She hesitated briefly as she walked past. "Afternoon, Mr. Miller. A good rain." Then she continued on her way, waddling with the weight of a pail of milk. He watched her, deciding that even Grace couldn't make carrying a heavy bucket look the least bit attractive. He returned to searching for someone to help, but there was no one else left in the barn. He stopped to pet Thunder, who had stuck his head out from his stall. Petting Thunder took him back to the moment that he met Grace. He laughed to himself as he thought of the contrast between the horse and the woman. The horse was gentle and calming, whereas Grace would probably never be a woman a man took comfort in.

"How you doing, boy?" Matt asked. "Glad to be out of the rain?" Thunder pushed at his petting hand, gently moving his head in different angles. The thought occurred to Matt that if he were stuck in 1762 he would need to learn to ride, and Thunder might be the perfect horse. Matt wasn't sure how long he stood there petting Thunder, but somewhere during that time the question that had been nagging him came to mind. "Dad would understand if I wanted to stay," he said to the horse.

Thunder reacted to his voice by picking his head up and making a *pwafft* sound.

"That's the best advice I've heard so far," Matt said. The horse answered him with big brown eyes.

**\*\*\*\*\*\*\*\*\***

The dog was already waiting in the barn out of the rain when he returned. Matt reached into the bin, grabbed his pack, and fished around the bottom for his pad of paper and a pencil. He marveled that both were in reasonable shape for having been under the weight of everything else he carried. He inspected the pencil in his hand with a new appreciation. He had used pencils his whole life, but only now did he truly look at one. As he rolled it between his fingers, he marveled at its simplicity. It was a plain number two pencil, painted yellow, with an eraser. However, from the 1762 perspective, it was no less than a techno-logical marvel. He saw no way that he could make a pencil without a great deal of time and skill. Almost every aspect of what made a pencil a pencil seemed hopelessly impossible to reproduce by hand.

Matt set the pad and pencil down and stood. He needed to do some brainstorming, so he walked to the corner of the barn, grabbed a giant corroded piece of tin and propped it on the bench. He picked up a large chunk of chalk that they used to mark the horses and wrote on the tin like a blackboard. He lectured his only student. "Well, dog," he said. "How would I make a living if I stayed in 1762?"

Matt wrote "INVENTIONS" at the top of the piece of tin and then started his list.

*1. Lightbulb*

"I could invent the lightbulb," Matt said. He thought about it for a moment and crossed it out. "Nowhere to plug it in."

*2. Cotton gin*

"I don't even know what a cotton gin does," Matt admitted. "Something about seeds." He crossed that out too.

*3. Steam engine*

He crossed it out without giving the dog any explanation at all.

*4. Radio*

*5. Television*

"There wouldn't be anything on worth watching," Matt said. Both got an X.

*6. Battery*

"I'm probably the only chemist in the world that can't make a proper battery," he admitted. "There has to be something I know how to do." He stepped back for a moment to ponder his expertise. "I got it," he exclaimed. It was loud enough to make the dog's ears stand up.

## 7. Drugs

Matt turned the piece of metal over and started to draw. When he was done, he stepped back to admire his handiwork. The chemical structures for ibuprofen and aspirin were written on the metal blackboard.

Common Name: Aspirin
Chemical Name: Acetylsalicylic acid
Natural Source: None

Common Name: Ibuprofen
Chemical Name: Isobutyl propanoic phenolic acid
Natural Source: None

Their session was interrupted by knocking on the barn door. "Are you in there?" Will called.

"Come in," Matt answered.

"Mother sent me to let you know we're having supper soon," Will said as he shut the door against the rain. He looked at Matt's chalk drawings and read the labels "Ibuprofen" and "Aspirin."

"Apothecary symbols," Matt said. Matt thought he might need to explain further, but Will's focus had already returned to dinner.

"Come over in a half hour," Will said. He was leaving when he spotted the pencil and paper. He picked them up and began writing. As with the chemical structures, Matt waited for Will's questions about the modern pencil or paper, but these too seemed to make no impression. When he was done writing, Will said, "For the meal. Let it be our secret."

Matt looked at the writing on the pad and gave him a thumbs up.

"See you soon," Will said. He walked out of the barn, sliding the door shut as he left. Matt took the note that Will had written and spent the next fifteen minutes practicing. He then filled the basin with water, washed his face, and brushed his teeth. He put on a clean shirt, smoothed his hair, hurried to the house, and was soon sitting with the family, ready for dinner. Grace walked in late, wet from the rain.

"Why are you all wet?" her mother asked.

"I was looking at Joshua," Grace replied. "The swelling and redness are much diminished and the open sores have scabbed."

Mary and Faith put bowls of food on the table. Grace followed them into the kitchen and returned with a steaming loaf of fresh-baked bread on a cutting board. The smell of the bread filled the room. "We should pray," Thomas said.

"I can," Matt replied.

Thomas had trouble hiding his surprise. "Are you sure, Mr. Miller?"

"I'd be honored," Matt replied. He paused and then bowed his head. Everyone followed, but he noticed out of the corner of his eye that Grace was watching him.

*Lord, bless this food and grant that we*
*May be thankful for Thy mercies be.*
*Teach us to know by whom we're fed;*
*Bless us with Christ, the living bread.*
*Amen.*

"A wonderful prayer, Mr. Miller," Thomas said. "Where did you learn this?"

"A friend taught me," Matt said, trying not to glance at Will. He looked over at Grace, who was calmly eating. She gave no indication that he had made any impression on her whatsoever.

"How was your first week of farming?" Thomas asked Matt.

"Difficult," Matt replied, "but I made it."

"My body has finally stopped aching," Will proclaimed.

"The quill has made you soft!" his father replied as he laughed heartily.

"All the thinking has made me hard up here, though," Will replied, pointing to his head.

"If we should want for any hard thinking," Grace said, "we'll let you know." The whole table laughed and Will nodded to his sister with a satisfied smile.

"What do you have planned for tomorrow, Mr. Miller?" David asked. "There's work."

"If it's all the same," Matt replied, "I want to go to Richmond."

"What will you do in Richmond?" Jeb asked.

"I'll check with the silversmith, talk to the apothecary, and then I have items I wish to buy."

"Like what?" Jonathan asked.

"Jonathan," Thomas said. "It's not polite to ask a man his business."

"That's okay," Matt said. "I wouldn't mind suggestions." Ever since his conversation with Thunder, the prospect that he might stay in 1762, invent things, and become a colonial gentleman had been gaining traction in his mind. He'd use the day in Richmond to explore the possibility. Matt began listing what he planned to do and buy. The most questions came about the clothes he would choose;

almost no one seemed to care whether he learned anything from the apothecary.

"What color waistcoat and breeches?" asked Jonathan.

"No idea," Matt said.

"You can't go wrong with blue or grey," Will said.

"Where's the best place to go?" Matt asked.

"I'd go to Henry Duncan," said David. "He has the best from London. The cheaper pieces are usually from last year, but—" He paused, and it was obvious that he was looking at Matt's black tee shirt. "'Twill work for your purpose."

"Tell Henry of our recommendation," Mary said, "and he'll extend a fair price."

"It's Henry Duncan, then," Matt declared. "What about shoes?"

"Obi Hunter," said Will.

"Obi's shoes never last," his father said. "They look good when you buy—"

"I wear his shoes this day," Will interrupted. "They are the highest quality."

"I'm not only looking for fashion," Matt said. "I want to be able to ride a horse."

Grace broke her silence. "If you're serious about riding, you should buy boots at McKissach's."

"Oh!" Thomas said. "He charges so much!"

"They are the best," Grace replied.

"I'll check out this McKissach's," Matt said. "How much for boots?"

"Twice that of shoes," David said.

"Which is?" Matt replied.

"A pair of McKissach's boots will run you almost three pounds," Will said. "They are probably worth it. The cost would not be lost on the ladies."

"Richmond ladies like boots?" Matt asked.

"Most assuredly!" Will answered back. "Remember Graine?"

"The one we met in town who spends money."

"She'd notice those boots," Will replied.

"Graine Martin will spend her relations into oblivion one day," Grace declared.

"But you must allow," Will said, "she always looks beautiful." The room went silent as Grace pondered her reply.

"I have often envied Graine Martin—"

A roar led by Will went up around the table.

"But wait," Grace said, laughing. "I've not finished."

"Dear sister, what could ever redeem your allowing you've envied Graine Martin?" Will asked, laughing.

"I have envied Graine Martin for her ability to persuade."

"How so?" Will asked. It was obvious he felt he had the upper hand in this sibling rivalry.

"In persuading her father 'twas worth financial ruin to impress Richmond and its Horse Prince."

"Touché, dear sister!" Will said. "You give as well as you get; not to mention that it's obvious you've been speaking to my fellows."

"Dear brother," Grace said, laughing, "I have not talked to your fellows at all."

"You didn't decide to call me Horse Prince from out of the heavens," Will said.

"One or two of the ladies your fellows are attempting to court may have mentioned it," Grace replied.

"They call you Horse Prince?" Jeb asked, looking at his brother.

"Prince of Horses, actually," Will replied.

"I hope you don't encourage that," Mary said.

"They joke," Will said. "How did this become about me? We were trying to figure out why Grace envied Graine Martin."

"I don't envy Graine Martin," Grace said with humor in her voice. "I often pity her when she comes to church wearing all those scarves and hats. Have you not seen?" Grace pantomimed a fanning woman and then said in falsetto, "I'm glowing."

"She *is* always fanning herself," said Jeb. "What's glowing?"

"Code ladies use for sweating," Matt said. "Though I'm no expert by any stretch."

"Ha!" David said. "Live a longer life and you'll learn that no one is. Ladies are magical creatures understood by no man."

"But how would you survive without the fairer sex?" Faith said.

"I could not, of course," said David. "You would allow, though, that there are men who would be better without."

"As there are ladies," Mary said.

The Taylor family was quite lively and it surprised Matt to discover that very few topics were off limits for their dinner conversation. "Back to my original question," Matt reminded everyone with a smile. "Boots or shoes?"

"A pair of each," Grace said, "especially if you plan to ride to Philadelphia."

"I thought you didn't know how to ride," Jeb said.

"That's another issue," Matt said. "Would one of you to teach me to ride a horse?"

The table went silent. This surprised Matt and he felt indignant that they judged his desire to learn. "I didn't think it was an unreasonable request," he declared. He waited impatiently for an answer, but the table remained quiet. Everyone had turned to look at Thomas.

"I should be the one to teach him," Grace said.

"You know you're not allowed to ride," her father said sternly. "You cannot teach him without riding yourself. Will should instruct Mr. Miller."

"Will doesn't know how to ride," Grace said. "Not properly."

"I prefer the carriage," Will explained.

"I spoke in haste," Grace said. "You ride as well as any man." The table went uncomfortably silent again. Matt made a conscious decision to keep his mouth shut. "Father," Grace said. Her voice was no louder than normal, but it was like a bomb going off considering the silence that preceded it. "I want your permission to ride."

"You cannot," he said.

"I should be able to ride the horses I care for every day."

"You cannot," Thomas repeated. "It must be Will that teaches Mr. Miller."

Will didn't hide the fact that he had no interest in lessons. "I could," he said, "but my sister would be the better."

"I forbid it," the older man said. He was less forceful this time. "I'll not let another daughter be taken."

"'Twas the Lord's will," Grace said. "No one was to blame."

"A lady doesn't belong on horseback," Thomas replied. "We'll speak no more of this."

Mary woke from her silence. "I respect my husband's decision," she said. "I confess, though, I've never regretted the part I had in allowing Kathryn to ride. 'Twas truly God's will."

Grace said, "Mother, I—"

"The decision has been made," Mary cautioned. Dinner ended quickly after that.

Matt took the opportunity to sneak back to the barn in time to receive a text. He sat there with the dog, watching the phone, wondering what new insight it would bring from the twenty-first century. A small voice inside him hoped they would inform him that rescue was impossible so he wouldn't have to agonize over what to do. Tonight's dinner with the Taylors reminded him that 1762 had its own challenges, but he still suspected he might use his advanced knowledge of science to accomplish great things. More than a small part of him was now looking forward to making some contribution to colonial America. His knowledge of smallpox alone could probably help thousands.

Though he was expecting it, the beeping of the text still startled him. He felt strangely like he did as a teenager when his girlfriend's father would walk into the room. He knew it had something to do with his current line of thinking, which ended with him informing his rescuers that their efforts had been in vain and he wasn't interested in coming home.

"Can you get back to the same place you arrived?" it read.

Matt typed, "Yes."

"We can open a portal for five minutes. You must step in. It will close and push you back."

Matt typed, "Is it safe?"

"We expect similar side effects."

Matt typed, "I might stay. Would rather not risk going into a wormhole."

"You can't stay. You'll change the future. Will text again in 24hr with details."

Matt typed, "How do you know the future doesn't require me to be here?"

There was no reply. Matt stared down at his phone as he talked to the dog. "They can't make me go back." He shut his phone off and stuck it in his pack.

# CHAPTER 21.

# APOTHECARY

---

Despite having to wake early for the ride into Richmond, Matt was in good spirits. He had finally been able to stand up after sleeping without his head exploding. After last night's texts, he had thought long and hard about the consequences of staying in the past. The whole concept of leaving the future behind and beginning again in colonial America filled him with both fear and exhilaration. He wasn't quite sure that he was willing to step out into the abyss just yet, but he would spend the day in town figuring out whether he could make it work.

David was quiet during the first part of the trip, so Matt took the opportunity to think of a cover story. He wanted to have answers to questions about where he came from, why he was dressed strangely, and why he knew nothing about eighteenth-century clothes. After practicing a few elaborate stories in his head, he decided it would be best to say as little as possible, mostly because keeping track of the lies was becoming difficult. Matt hoped Henry Duncan would be satisfied with his request for clothes and the fact that he had been recommended by the Taylors.

David interrupted his thoughts. "Who do you think should give the riding lessons?"

"It's up to Thomas," Matt replied. He really did believe this.

"I don't endeavor to meddle."

"No one ever does—" Matt said. It came out more harshly than he wanted. "Sorry, that's not the way I wanted to say that."

"You're forgiven," David said, "but I understand the intent."

"My mother and father live in different cities," Matt explained. "I wish one of them had half his conviction."

"It must have been difficult growing up without a family."

"You get used to it."

"I agree that a family wants for a leader," David explained. "Thomas and I grew up with a very strict father. There should be a place in between."

"I don't want to get involved," Matt said. "I've got enough to—"

"You started this."

"Not on purpose!"

"Will you have a family?"

"Of course."

"How would you deal with your daughter's request to ride?"

Matt answered reluctantly. "If it were me," he said, "I'd let her ride. You can't expect to protect her from everything."

There was a long pause before David spoke again. Matt's answer seemed to satisfy him. "I'd go to first to Henry Duncan's," he said. "Have him hold the packages so you don't have to carry them in the rain. We can retrieve your suits on the way home."

"How much will it cost me for two changes of clothes?" Matt asked. He thought twice about his question and then added after a pause, "In Richmond."

David looked at him out of the corner of his eye suspiciously. "Not much different than in Philadelphia, I imagine."

Matt replied with a blank stare.

Eventually, David rolled his eyes in resignation. "Two changes shouldn't cost you more than four pounds, including a good coat. Buy two pairs of breeches, two pairs of stockings, and maybe three shirts."

Matt nodded.

"I'll point you in the direction of the apothecary when I drop you at Henry's shop," David said. "Benjamin Scott... between us...he's an odd fish."

"I'm not expecting much," Matt said. "Anything he could tell me would help."

David laughed. "We'll see how you feel later."

"I'll stop at the silversmith after and see if he's sold my ring," Matt said. "Then I want to check out a general supply store."

"If you can wait until midday," David said, "you can come with me to Adam's."

"Great," Matt replied. They passed a few buildings upon entering town and then stopped in front of Duncan's Clothier. Matt made plans with David to be picked up in front of Samuel Smith's at midday, and then hopped down from the wagon.

*********

A bell echoed through Duncan's shop when Matt opened the door. It was bright, organized, and clean, and

Matt was immediately impressed. A well-dressed man in a grey wig stepped from the back room and exclaimed, "Henry Duncan!" as he walked out to shake Matt's hand. "How can I help?"

Henry stepped back and looked Matt up and down. He stared long and hard at his zip-off hiking pants and even more closely at his high-tech athletic shirt. "I can't even imagine where you got those," he proclaimed.

"That's easy," Matt replied, smiling. "I spent the last couple of years in the Orient. It's all from China." It tickled Matt to no end that all his clothes had "Made in China" labels.

Henry reached out and felt Matt's shirt between his fingers. "That's the oddest-looking silk I've ever seen."

"You can buy some strange things in China," Matt explained. "Thomas Taylor sent me. I'd like your help picking out two suits of clothes."

"I simply love the Taylors," Henry replied. "What kind of clothing?"

"Something to ride a horse," Matt explained, "and also something a little nicer, like what you would wear to church."

"You probably want to impress the ladies, then," Henry said.

"One lady, anyway," Matt replied without thinking much about it.

"Not that beautiful Grace Taylor?"

Matt opened his mouth to say something, but nothing came out.

"Your secret's safe with me," Henry said. "You know she's refused the hands of numerous Richmond gentlemen." Matt was still trying to figure out how he'd become so easy to read. Henry was now inspecting his face. "You might be handsome enough," he said. "Just barely." He put a hand on his chin as he continued his thought. "I'll do what I can to make you a Virginia gentleman."

"I'm on a budget," Matt said. "I told you that, right?"

"As are all young gentlemen," Henry replied. "An aspect of my trade I much appreciate."

This surprised Matt, who said, "I'd think you'd want rich old men in here."

"It's not all about silver, my boy!" Henry exclaimed. "Young men hope to impress. I enjoy working within their means." He paused for effect. "I take their money, but I make sure not to empty their pockets."

*********

Matt walked out of Henry Duncan's soon after. Despite spending more than he had planned, he was pleased. He had a reasonable stack of clothes and had received instructions on how it all should be worn. This included a thick coat that Henry had agreed to sell at a considerable discount, but only after Matt guaranteed he'd still be pleased with it after realizing it was two years out of fashion. Henry had also recommended a shoe shop nearby that, in his words, was "absolutely fabulous," which made Matt smile.

After considering how much he had spent on clothes, Matt decided not to make a trip to McKissach's for boots. Shoes made more sense for now. He'd think about boots when he was sure of his financial status. There seemed to be a shoe store on every corner in Richmond, but he eventually found Fullerton's Shoes, the shop that Henry had recommended. Matt walked out the door in less than thirty minutes with a pair of buckled black shoes in a burlap sack.

He proceeded to the silversmith, passing a number of shoppers carrying packages or baskets. He touched his hat to them when he could and said "Good day" like he had seen Will do. The sign on the door of the silversmith's said "OPEN," so Matt walked in without knocking. He was anxious for an update on the prospects of selling the ring. His decision to stay in 1762 might be easier to make if he knew he wouldn't have to struggle from the very beginning. He at least needed enough money to buy a horse. Jacob Berkley was talking with a customer, but acknowledged Matt as he walked in and said, "Feel free to look around."

Matt meandered around the shop to look at the various pewter and silver pieces on shelves lit by skylights. Some of the better-lit items reflected the sun and clouds, and he could see his own reflection in the ones that were most brilliant. He rubbed his chin, noticing that he needed a shave.

"Mr. Miller, we can speak now," Berkley called.

Matt turned to shake Berkley's hand. "Good day," he said. "I was in town and thought I'd stop for an update on the ring."

"Certainly," Berkley replied. "It's too early to have particulars, but the prospects look good. I've sent it to Maryland for auction and also given notice to four associates in Philadelphia, two of whom would surely be interested. Our position would be greatly enhanced should they attend."

"You sent it to Maryland?" Matt asked, concerned.

"Protected courier," Berkley replied simply.

"How much do you think it's worth?"

"There's a reserve of eighty pounds," Berkley replied. "The auctioneer won't sell it for less. We can expect near three hundred pounds with the proper patrons."

"Three hundred pounds?"

"I know," Berkley exclaimed. "I should have given you sixty at the start."

"Or I should have taken it to Pennsylvania and sold it myself," Matt said, trying to sound incredulous.

"You could never have sold it for more than a hundred," Berkley said. "And there's always the chance that bandits would've taken it from you on your journey. You made the correct decision."

"I jest," Matt said. "I'm fine with our deal."

"The auction is late next week," Berkley said. "Should it sell, 'twill take another week for funds to make their way to Richmond."

"The best of luck to both of us, then," Matt said. Matt reached to shake Berkley's hand and said goodbye. He walked out of the shop and headed toward the apothecary.

*********

Benjamin Scott was a disheveled man sitting behind a counter reading a scientific text next to a glass jar of leeches. Medicines surrounded him and were stacked on every shelf. "Good morning," he said. "How may I help you?"

"I was hoping to buy a few supplies and ask some questions," Matt said.

Scott shut his book to indicate that Matt had his full attention. "Always have time to talk to a customer," he said. "Are you ill?"

"Not at all."

"Then why are you here?" Scott said, disappointed.

"I want to start my own apothecary in Philadelphia," Matt said. "I've been away in the Orient and am unaware of the current state of the business."

"Where did you apprentice?" Scott asked.

"China."

"What in God's name do they know about apothecary in China?" Scott replied. It was more commentary than question.

Matt pretended he hadn't heard him. He was already thinking about a plan to make either aspirin or ibuprofen if he decided to stay in the colonies. He looked up on the shelf. "Do you have willow bark?"

"For what?"

"I can change the willow bark so it doesn't upset your stomach," Matt replied.

The older man became more attentive. "You can improve upon willow bark?"

"Once I get the ingredients used by my Chinese master," Matt said.

"Which are?"

"Willow bark is mixed with something called acetic anhydride," Matt said. "The reaction will work with distilled vinegar, but not as well."

Common Name: Salicylic acid
Natural Source: Willow tree bark

Common Name: Acetic anhydride
Natural Source: Unknown

Common Name: Aspirin
Natural Source: None

"I don't have this acetic anhydride," Scott said, "but I do have the vinegar." He reached up for a blown-glass bottle marked "Distillate of Vinegar" and placed it on the counter.

"That's a start," Matt said. "Do you have anything called caustic soda or lye?"

"You can get lye from the soap maker," Scott answered, grimacing as if he was uncomfortable standing. "God-forsaken rheumatism in my hips and legs," he explained. "Standing causes some annoyance." He took a seat and motioned for Matt to sit.

"Do you take anything?" Matt asked.

"Willow bark, coincidentally," Scott replied. "I've not had any as of late due to my dyspepsia." He patted his stomach to emphasize the upset.

Matt pulled out a sachet of paper from his pocket that he had made containing four ibuprofen tablets. He was hoping Scott would be convinced to help him once he felt the effects of ibuprofen for himself. "The medicine I spoke of is here," he said. He showed Scott the tablets. "It's very rare and expensive. Swallow two and your rheumatism will go away for the rest of the day."

Scott gazed at him suspiciously. "What makes you think I'd swallow anything a stranger brings in here?"

"I'm looking for advice," Matt said. "It wouldn't help me at all to make you sick."

"I didn't say you'd poison me intentionally," Scott replied. "I'm not convinced you know your trade."

"I only have four tablets," Matt said. "I'll swallow one in front of you to prove it's not poison, but that's one less to make you feel better. You swallow these with water without chewing."

"Let me see," Scott said, holding out his hand and taking the opened sachet. "These were manufactured on a press of sorts. These will cure my pain?"

"Lasting five hours or so," Matt replied. "You'll move easier for the rest of the day."

Scott eased himself to his feet, poured two glasses of water from a metal pitcher, limped back, and set them on the counter. "You first," he said.

"You need two for full relief from rheumatism," Matt replied. "I'll swallow one so you have a half dose for tomorrow." He picked up a tablet and glass of water and was about to pop it in his mouth when the man grabbed his hand.

"I trust you," Scott said. "I'll take two today and two tomorrow."

"You won't regret it." Matt put the tablet back into the sachet and pushed all four across the counter. Scott swallowed two easily.

"You should start to feel better in about fifteen minutes," Matt said. "It takes a couple of hours to loosen your joints."

"Why are you giving me your rare medicine?" Scott said.

"For your time," Matt replied. "I want to work with someone who knows the apothecary trade."

"I'm not taking an assistant," Scott said, "if you're looking for work."

"I need a place to make this medicine," Matt said. "Do you have a lab?"

"In back," Scott said. "I'm not sure I'd want to share it."

"Even for a fee?" Matt asked. "I'll guarantee you a supply of this medicine."

Scott thought for a moment. "Two shillings per week to use the room, and I'll want two pounds before," he said. "I'll return the two pounds if you leave it clean."

"Two pounds!" Matt exclaimed.

"I'd want it cleaner than when you started," Scott replied, unfazed.

Matt finally nodded in agreement. "I have other questions."

"We'll talk again when our arrangement is formalized," Scott said. "A man doesn't give away a lifetime of experience for free."

"That's it?" Matt asked.

Scott shrugged his shoulders. "I'm starting to feel better already," he said.

Matt handed him a list. "Would you try to get these things? I'm willing to pay a fair price."

"I'll search," Scott replied.

# CHAPTER 22.

# LEVI

---

Matt stood out on the street outside the apothecary, trying to see the sun between the rain clouds. There was a light drizzle in the air. He decided he probably had some time before David would arrive, so scanned the block and located a dry bench across the street covered by a tree. It looked like the perfect place to sit, rest his feet, and make plans.

The street was crowded, and Matt had to step between two men as he walked out to cross to the bench. He was forced backward as a speeding wagon, seemingly on the wrong side of the street, veered away from him to avoid a collision. In the driver's attempt to turn away, the horses ran directly into a stack of empty wooden barrels set on the curb by one of the merchants. The larger of the two drivers pulled up on the reins and stopped the wagon abruptly, but not until one of the barrels was lodged between the two horses under their harness. The barrel needed to be removed before the wagon could move forward.

Aware of his part in causing the accident, Matt trotted to the wagon. He called out, "Is everyone okay?"

"Why the hell don't you watch where you're going!" the driver said. "You!" The man's face was a dark shadow silhouetted against the midday sky.

Matt squinted up and said, "Hey, I'm sorry." He met the man's gaze and finally recognized the face; it was Levi Payne. Matt couldn't believe his bad fortune.

"Look what you've done to my horses," Levi shouted. Matt glanced over at the animals, which had stopped in their tracks after knocking two barrels into trees past the curb. Despite the barrel stuck between them, neither horse looked injured. Both seemed to have already recovered from the accident and contentedly nibbled on the leaves of the tree hanging over the scattered barrels.

"The horses look okay," Matt said. "I apologize again." The wagon had seemed to come out of nowhere and he had been so surprised that he didn't really know how much had been his fault.

"Idiots like you shouldn't be walking our streets," Levi said. "I thought you were going back North."

"Levi," Matt said slowly and assertively. "I'm sorry. It was an accident. I didn't intend to walk in front of your speeding wagon."

"You think sorry's good enough?" Levi replied.

Matt had no desire to fight in the street. "Sorry is the best I'll be able to do. There's no damage. I'll help clean this up and we can both be on our way." He glanced around at the townspeople who had come to see the commotion. Most had paused to watch from a safe distance. One man walked to the other side of the street, seemingly after recognizing who was driving the wagon. Matt looked back up to try again to reach an agreement with

Levi, but was surprised to discover that only one rider, a black man, remained in the driver's seat.

Matt saw the punch coming in time to back away, but it still caught him across the front of his face. He dropped the sack containing his shoes and ducked, but Levi followed and punched him hard in his side. Pain shot through Matt's ribs like an explosion and he went sprawling. He regained his feet, staggering as Levi came charging. Matt's sidestep was automatic as he ducked to avoid Levi's swinging fist. He caught Levi as he turned toward him and was able to drive his fist straight into Levi's stomach. The blow was solid. Matt was satisfied that it was about as precise a strike as he could have made. His whole arm shuddered from the concussion like he had punched stone. The larger man staggered back, surprised at the strength of the blow, but was still on his feet. Matt had expected him to drop and had relaxed his posture. He saw Levi standing and resumed a fighting stance.

"I'll kill you, you son of a bitch," Levi said as he closed the distance between them. Matt timed his approach, braced himself with his back foot, and hit Levi in the chest with a sidekick. Levi dodged in time to avoid the full force of the kick. Matt realized that his timing was off and his flexibility was poor from not working out for so long. Still, the kick had hurt, so Levi stepped back to shake off the blow and reassess his opponent. Matt resumed his fighting stance with his fists in the air.

The tae kwon do fighting stance Matt was using looked like that of a boxer. His body was turned to the side to protect his vital organs from a frontal attack and to set up his back leg for kicks. His hands, as generally in tae kwon do, were up for protection or counterblows. He wouldn't initiate attacks with his hands, since it would be prudent to keep the man beyond arm's length for his kicks to offer any advantage. Once one of them closed the distance, it would become a brawl and the toughest puncher would win. Matt was now sure it wasn't him.

Levi circled to his side and Matt adjusted his body accordingly. As Levi advanced, Matt spun into a round-house kick. The kick would've been deadly if it was high enough to strike Levi's head squarely, but again his timing was wrong and his foot glanced hard off Levi's shoulder. Levi's forward momentum got him close enough to hit Matt with a forearm across his back. The blow pushed Matt into Levi's other fist, which drove into his chest, and then Levi hit him twice in rapid succession.

Matt's defense couldn't have been slower. He was paying the price for timing and reactions developed over years of sparring using body armor and under ring rules that didn't prepare him for the all-out pandemonium of a street brawl. He needed to back away and reset his stance. Levi was holding him and punching while Matt tried his best to untangle himself. Matt finally saw his opportunity and reached up to grab Levi's shirt collar, pulling his neck down and bringing his right fist solidly into Levi's face. Blood spurted from the man's mouth and nose as he staggered back and away.

Matt, now free from the barrage of blows, sucked in his first breath and pain seared through his chest as it expanded. Levi seemed strangely calm as he wiped the blood from his face. He glared intensely at Matt and said, "You're done." Matt resumed his fighting stance, taking deep breaths and trying to ignore the pain in his chest. One or two of his ribs was bruised or broken.

Matt looked back at Levi from behind raised fists, turned his front palm to the sky, cupped his fingers twice in a backwards-waving motion and mouthed the words "Come on." Levi walked forward. Matt timed his attack to Levi's approach, hitting him with a spinning sidekick that connected with the bone and cartilage above his left hip. The man yelled in pain but continued rapidly forward to reach Matt before he could set his feet firmly back on the ground. Levi's fist hammered into the side of Matt's head, causing his vision to go haywire as if something had shaken loose in his head. He saw the same flashes of light that he had seen in his dreams, but interrupted by images of his attacker. His vision sputtered like a high-speed camera.

Unable to see, and moving mostly on instinct, Matt thrust his knee up into where he guessed Levi's groin would be. He connected with something unseen and then fell backward onto the ground to more pain. Levi staggered away from the blow and smashed into the wagon. A gash opened on his head as he fell between the wheels. He struggled to pull himself up, looking back at Matt, who was still lying on the ground half-blind and unable to regain his balance. Levi limped quickly, holding his hip and wiping the blood from his face. He pulled his leg back and kicked Matt as hard as he could, despite the pain from his injured hip. The kick connected, but only mildly, and Matt rolled away. Levi limped forward.

He pulled his foot back to kick Matt again, but his leg gave way, and he fell to the ground. Matt pulled himself to his hands and knees to face his attacker, who was also attempting to stand.

"What the hell?" David had finally arrived.

# CHAPTER 23.

## PHYSICIAN, HEAL THYSELF

---

Matt had regained his feet and could almost see. Levi was limping forward saying, "I'll kill you."

Matt raised his front hand and waved him forward as he had done before. "Come try," he replied.

David jumped from the wagon holding a pick handle. "Levi, if you step any closer, I'll cave your head in." He waved the pick handle towards him threateningly. "You've had enough, both of you."

Levi wiped the blood from his face and glared at Matt. "You have your reinforcements," he said. "Next time I find you in Richmond, you'll not be so lucky."

"Bring it on, you ugly ape," Matt said as he raised his fists. "Better get used to me, because I'm staying right here." Matt liked the sound of his words as they left his mouth and he felt a strange, excited calm wash over him. *I've made my decision!*

Levi smiled, turned around, and limped back to his horses. The black man who had been holding the reins watched with a blank expression as Levi struggled to make his way to the wagon. When he got there, Levi said,

"Help me!" The man reached down to take his bloody hand and pulled him up to the seat. Levi tugged on the reins, backing the horses off the barrel, and shook them to guide the animals around the obstruction. The barrel glanced off the moving wagon wheel and tumbled out into the middle of the street. Levi didn't look back as the wagon sped away.

"You all right?" David asked.

"I feel like I was hit by a freight train," Matt replied.

"A what?"

"Never mind," Matt said. "I got hammered."

"He looked worse than you!"

"That crazy SOB would've kept coming until I was dead," Matt exclaimed.

"You both lost this battle."

"I might need help getting into the wagon," Matt said. He took one step and collapsed unconscious onto the ground.

**\*\*\*\*\*\*\*\***

When Matt awoke, drops of water were falling on his face and he was staring up into the cloudy sky in the back of a moving wagon. He could hear the pattering of rain on the canopy, which fell in time with the drops hitting his face. He pulled himself to a sitting position, then to his knees, and stepped over supplies as he went to the front and climbed over the seat back. He dropped in beside David.

"How do you feel?" David asked.

"Not too bad," Matt replied. He took an inventory. "It aches when I breathe, but everything seems to work. All my teeth are here. My hands are bleeding." A mixture of mud, sweat, and blood covered his arms.

"Your face looks horrible," David said.

"It doesn't feel that bad."

"We're almost there," David declared. "There's a jug of water." He pointed to the floor of the wagon. "We'll get you victuals. I had intended for us to eat in town."

"How long have I been out?"

"Couple hours," David replied. "It took a few people to lift you into the wagon. I visited the general store while you were out."

"Ah!" Matt exclaimed. "There were supplies I needed."

"You would've been in no condition to walk," David replied matter-of-factly.

"I forgot my clothes," Matt said.

"I got them," David replied. "Make sure the bleeding has stopped before you put them on."

"Bleeding?"

"On your back. Of all people, why did you pick a fight with Levi Payne?"

"I'm paying for being a smartass the first time we met."

"Smartass?" David said. "That's a peculiar term."

"Smartass," Matt repeated.

"You're in Richmond little more than a week and you've quarreled with him twice?"

"Sorry to spread the blame around, but I'm sure this had something to do with me working on your farm."

"You called him an ugly ape," David said, laughing.

"I don't remember," Matt said.

"He is an ugly ape. Capital watching him bleed. I didn't predict you for a fighter, though."

"I'm not a fighter," Matt said. "I stepped out onto the street and he almost ran me down with his wagon. I apologized twice trying to keep it from getting out of hand." They were moving through the gates to the farm.

"Maybe you should have apologized three times," David said. "Have Mary clean your wounds."

"I can wash at the well."

"No," David said. "You don't want those cuts to fester. I insist."

David stopped the wagon at the side of the farmhouse. As Matt stood up, his head thundered and pain shot through his ribs. He held onto the side of the wagon and lowered himself slowly to the ground, then limped toward the kitchen. The kitchen door opened and Grace stepped through with an armload of pots. She looked at Matt with a stunned look on her face. "You get run over by a wagon train?"

Matt laughed and nearly doubled over in pain. "Something like that," he said, gasping for air between the laughs. "Trouble in town. Would you mind getting your mother? I heard she might be a good nurse."

Grace set the pots down and stepped to help Matt into the house. She held the side of his arm with surprising strength and lifted him as well as any man her size might have. She supported him as they walked into the kitchen, and then helped lower him onto a chair.

"Don't go away," she said with a laugh. "I'll get Mother." She stood there then, looking at his bruises, paralyzed.

"Grace," Matt said, "can you get your mother?"

"Yes," she replied. "Sorry." She walked past him and out the door. He wasn't sitting longer than a minute when Grace walked back into the house with her mother.

"Explain yourself, Mr. Miller," Mary said. "You've been fighting."

"I'm in too much pain to explain myself," Matt replied. "I told David I should do this on my own." He started to stand.

Grace put her hand on his shoulder and forced him back in his seat, surprising him again with her strength. "Sit down and don't be a baby."

"Well?" Mary asked.

"I got in a fight with Levi Payne."

"Did you fight back?" asked Grace.

"I got this way begging for mercy," Matt said. The breath he took to say "mercy" sent a shocking pain through his ribs.

"I hope you hit him at least once," Grace said. "He's an animal."

"Maybe once," Matt said. "I'm about ready to fall off this chair. Either help me or let me leave."

"Don't leave," Mary said. "We'll return in a moment."

It seemed an eternity before they came back. Mary had an armload of towels and soap, and Grace was carrying two buckets of water that she took into the kitchen to exchange for water already heating over the fire. She returned with a bucket of steaming water and set it down beside the table. Mary looked at Matt's wounds and then at Grace. "You should learn this."

"Learn what?" Grace asked.

"To clean wounds so they don't fester. If they aren't cleaned while the blood still flows, the wounds become poisoned. It should be boiled water. In Germany they'd say that there are unseen beasts within the water that cause this poison. It's the same when you assist in childbirth."

Matt felt the need to speak up. "The best way to prevent the poisoning is to boil everything and to make sure your hands have been washed with plenty of soap before touching the wounds or delivering the baby." He subconsciously glanced up at Grace at the mention of hand washing. She met his gaze.

"The back of his shirt is covered with blood," Grace said.

"It must be removed," said Mary. She looked at her daughter, paused, and said in a commanding German accent, "Grace, it would be improper for an unmarried lady to see a man so."

"Beat up and bloody?" Matt said, joking through painful breaths. Mary ignored him.

"I'll do this for my own sons someday," Grace said. She motioned towards Matt. "I might as well practice on him."

"Fine," Mary said, resigned. "Mr. Miller, remove your shirt."

Matt attempted to take it off himself, but as soon as he tried to lift his arms, the pain in his ribs forced them down. "I need help," he said.

Mary eased off Matt's shirt as he leaned forward in the chair. He grimaced in pain. Grace gasped. He followed her eyes to the reddish bruises that covered his rib cage. Matt said, "It looks worse than it feels."

"I hope so," she replied.

Mary pulled a wet, steaming towel from the water. "This will sting," she said.

"Go for it," Matt said. Both Grace and Mary looked at him strangely.

"Go where?" Grace asked.

"Do it," Matt said. Mary instantly took action; the dripping towel was on him and he couldn't believe the pain. "Ouch!"

"As my daughter said, don't be a baby."

It took about ten minutes to clean his wounds with soap and water. He was now sitting in a puddle and his pants were soaked. Mary pushed him forward to look at his back. "How bad is the cut?" he asked.

"It should be sewn," Mary said.

"Sewn?" Matt exclaimed. "With what?"

"A needle and thread," Mary answered. "The injury is too large to heal."

"Are you sure?" Matt said. He saw Grace smiling. "Don't even say it," Matt exclaimed. "I'm not a baby. You're going to put a needle and thread through my back without any sort of painkiller."

"Do you desire your rum to get drunk first?" Grace replied.

"I don't—" Matt caught himself midsentence. He gave her a dirty look. "I'm too tired and hungry. Just do it."

# CHAPTER 24.

# REVELATIONS

Sewing up his back took another twenty minutes and was as painful as he had imagined. He limped back to the barn when they'd finished, holding his bloody shirt. His legs were fine, but the top part of his body felt delicate, like he was made of glass. His pants were uncomfortably wet. All in all, he imagined himself one big, bleeding wound. He couldn't remember the last time he had been in a real fight; maybe not since the elementary school playground.

Matt reached up to slide the barn door open and pain shot through his side. He saw that David had placed his parcels on the workbench. He had hoped to try the clothes on after he got home, but now he couldn't touch any of it until he healed. Even if he wasn't bleeding, he was in too much pain to perform the acrobatics required to change in and out of the colonial clothing.

Matt forced himself to walk to the well and fill the wash basin, brought it back in, and then started the painful process of untying his shoes, removing his pants, and washing his lower body. He dried himself and put on a fresh pair of underwear and pants, looked at his new shoes and decided to at least give them a try. His feet slid in easily and he fastened the buckles. Just as he was standing to walk, there was a knock on the door.

"You look horrible," Will said.

"You should have seen the other guy," Matt replied. He reached into his pack to grab a shirt, which he slowly slid over his body to hide the bruises.

"How did it start?"

Matt sat and told the story as best as he could remember. "And then he was walking towards me like a bloody zombie," Matt said, finishing the story. "One of us would've ended up dead if David hadn't interrupted."

"What's a zombie?" Will asked.

Matt instantly wished the word had never left his mouth. He struggled to come up with some reasonable definition. "It's from a children's story they tell in Philadelphia," he explained. "A zombie is a corpse that comes back to life and walks around kind of slow and tries to eat people."

"That's a children's story in Philadelphia?"

"The point is that the man was moving toward me like a bloody walking corpse and wasn't giving up until I was dead," Matt said.

"David said that Levi could barely walk."

"Maybe."

"I'm anxious to see how much damage you caused," Will said. "I hope you're well enough for the King's Tavern on the Sabbath."

"Don't the Paynes go there on Sunday?"

"I guess they do," Will said with a clever smile. "We'll invite Father. Business as usual."

"That's almost a whole day and a half away," Matt said sarcastically. "I'll be better by then."

"I thought you'd want to come," Will replied. He changed the subject. "Mother said supper will be in about half an hour. There's cold, fresh water."

"I'll be there," Matt replied.

Will turned around and left, shutting the barn door behind him. Matt almost followed him, but thought better of it. The barn was quiet and he wanted to sit and rest, so he walked to his bench and sat with no other purpose than to sit. A strong sense of satisfaction was starting to replace his initial shock. He had stepped into the abyss and come out the other side, still alive. He looked up at the roof and thanked God, then popped to his feet and walked to the farmhouse.

Like always, the house was a flurry of activity centered on serving the meal. Will and his father were deep in conversation. Thomas waved him over. "Have a seat," he said, pointing to a chair. "You had some excitement."

"Sure did," Matt agreed. "I bought two full changes of clothes, shoes, and met the apothecary."

"You know I mean the business with Levi," Thomas said, smiling.

"Oh, that?" Matt made a waving motion with his hand. "No big deal. Men try to beat me to death almost every day in Philadelphia."

"I can't imagine he'd have killed you," Thomas said.

"He said he would," Matt replied.

"It's the chance you take when you fight," Thomas said.

"I tried everything I could to avoid fighting, but I wasn't going to let Levi Payne pound me into a bloody pulp."

"If what David says is true," Thomas said, "you have some experience as a fighter."

"Do I look like I get into fights?" Matt said, motioning to his face. He realized his mistake simultaneously with the laughter that erupted from his two counterparts. "So maybe I look like I fight now," he corrected, "but I'm not used to fighting...really."

"A man watching the fight said you kicked Levi in his chest," Thomas said.

"You kicked him?" asked Will. "How?"

"Something I learned in China," Matt said. "I planted my back foot and kicked."

"So you know how to fight," Thomas said, "but you pretend not to fight."

"Hard to explain," Matt said. "I did it for exercise."

"You fight for exercise?" Will asked, surprised.

"You wear armor to protect your body and head," Matt said.

"Armor?" Will exclaimed. "Like a knight?"

"No," Matt replied, laughing. "It's made of thick leather that goes around your chest to protect your midsection." He pointed to his stomach. Then he pointed up to his head and said, "You wear a leather helmet for protection from head kicks."

"People can kick you in the head?" asked Will.

"Yeah," Matt said, nodding. "I tried to kick Levi in the head when I was certain he intended to kill me. I couldn't get my leg up high enough and he used the opportunity to crush *my* skull." Matt pointed to a big bruise on the side of his face. "My vision went crazy. I got lucky to catch him with my knee."

"Can you teach me this kicking?" Will asked.

"It's something that I practiced for years," Matt replied. "It would take a long time to get it right. Someone like Levi would grab your leg and break it in half."

"Still," Will said, "I desire to learn this."

"When I feel better," Matt said.

"Supper is ready," Mary called from the table as she set a loaf of bread down. Everyone got up and went to take seats at the table. Matt noticed Grace looking at his bruises while everyone was passing the food. Jeb said the prayer and the family began the meal in their typical roar.

Jonathan spoke as he was inspecting Matt's bruised face. "Levi Payne dislikes everyone on the Taylor farm," he said.

"I don't think he dislikes everyone," Thomas replied.

"He fought Will," Jonathan said, "and he dislikes Grace because she wouldn't marry him."

"That's a private matter," Mary said.

Jonathan looked at Grace and said, "Sorry, Grace."

"'Tis fine, Jonathan," Grace said. "I don't see why it shouldn't be known. I refused to let Levi Payne court me." She looked again at Matt's bruised face. "Must we fear him beating Jeb next because he couldn't court me?"

"He didn't beat me," Matt said.

Grace gave him a dirty look and hissed, "You're missing the point."

"You think he wants to fight me too?" asked Jeb.

"See what you started?" Mary said.

"I apologize," Grace said, "but it could be any one of us, or those who work for us." She stared at Matt. "It's not boys fighting anymore. Is this all because of me?"

"Not only you," her father said. "My son is correct. Levi has an aversion to our farm."

"He called Will a slick dandy," Jonathan said.

"How'd you remember that?" Will said, laughing at his little brother.

"Your face was bloody," Jonathan said, "and it scared me."

Will fingered a scar on his face. "I gave as good as I got," he said proudly.

"Would he threaten the men who work for us?" Mary asked. She turned to Thomas, who seemed to be thinking hard for something appropriate to say.

"He could," Thomas said. "This has been coming."

"What's been coming?" Matt asked.

"Nathan expected that we would lie down and die years ago," Thomas explained. "We've sold more horses than ever this year; two this week for one hundred and fifteen pounds."

"The twins!" Jonathan said. "Grace made them shine. Mr. McKinley kept walking around saying how beautiful they were. Just when you thought he'd finish, he'd step away and do it again."

"One time he gazed into the sky and thanked God for their beauty," Grace said.

"As we should," her father replied. "Tell us what you do to make the horses shine."

"They know," Grace said.

"Not Mr. Miller."

She looked at Matt and began to explain. "I brush them the day before they are to be sold; it takes almost an hour with the different brushes," she said. "This removes the dust and brings the oils from their skin. I change their straw so they smell fresh. When morning comes, I repeat it all to work the natural oil from the skin and then wipe them down with a damp cloth to give them a dark shine. Their coat becomes so full with oil that it darkens the hair to look wet. My hands and arms ached from getting those two ready."

"Would your time not be better spent on unsold horses?" her father said. He had a critical tone in his voice.

Grace looked at him, puzzled. "I take pride in my animals! I want our patrons to rush home to show their fellows. I was proud of—"

"Exactly!" Thomas exclaimed. "And so was I." He waited for a moment and added, "That's the difference between our farm and Nathan's. One hundred and fifteen pounds for two horses is a king's ransom. We don't sell to kings; we sell to townsfolk. Can you think of any pair of horses Nathan has on his farm that could sell for one hundred and fifteen pounds?"

"No," Will piped up. "They have naught these days."

"'Tis why Levi is frustrated," Thomas said. "His slaves care naught for his horses and whether they shine."

Matt had wondered about Thomas's perspective on slaves ever since he had learned that there were none on the farm. This seemed like an excellent opportunity. He knew to tread lightly and avoid questions that might back the man into a corner. "We don't own slaves where I come from," Matt said. "Is quality the only reason you don't keep slaves?"

"I made the decision long ago to conduct my affairs thus," Thomas said. He thought for a moment. "I stopped keeping slaves because they clouded my judgment."

"Judgment?" Matt asked.

"Slaves are property," Thomas replied, "to be bought and sold."

"You don't disagree that men should own slaves?" Matt said.

"I have no interest in keeping other men from owning slaves," Thomas replied. "I merely choose not to own them myself."

"I'm still curious why a farmer would make that decision," Matt said. He tried his best to keep an even and unbiased tone.

"Do you support a man's right to own slaves?" Thomas asked, looking straight at Matt.

"No," Matt replied, "though I imagine that voicing this opinion would be an easy way to be run out of Richmond."

"I'd ask that you not discuss slavery with anyone outside this family as long as you're my guest," Thomas said. "Naught would come from this transaction."

"I'm still curious how slaves caused you to divide up your business," Matt said.

"It was not in my character to break up families or buy and sell the young ones," Thomas explained. "I couldn't sleep with all the moaning and weeping."

Matt could see that the topic was starting to make the man uncomfortable. "Thanks," he said. "I appreciate your perspective."

"We've already spoken of this more than you can imagine," David said. "Many times we've weighed the advantage of slaves against the grief they bring. We still end in the same place."

"And that is?" Matt asked.

"Purchasing slaves might make sense next year," David said with irony in his voice.

"And next year and next year," Will said. "'Tis thus since I was a young boy."

"There may never be a proper occasion for slaves on this farm," Thomas said.

"Might we debate something else?" Grace said. "We don't need them. The men this season have labored long and hard and we should celebrate their hard work."

"Uncle selected fine fellows," Will said. He looked over at David with a smile.

"Many are quite friendly," Thomas said. "It seemed much too quiet today without them."

"Not for Mr. Miller," Jonathan said.

"I slept in the wagon," Matt replied. "That counts as quiet."

"What will you do for excitement tomorrow?" Grace asked.

"I was hoping to start my riding lessons sometime," he replied, surprising even himself. The whole table went quiet. Matt saw David trying to hide the wide grin he had on his face. Matt turned his attention to Grace, who was looking directly at her father.

"Do you think you'll be in a condition to ride?" Thomas said.

"Probably," Matt replied, trying not to look at David.

"You could take some of that ibuprofen," Jonathan said.

"How did you know that it was called ibuprofen?" Matt said, amazed.

"I listen," Jonathan replied.

"It'll work to relieve some of this swelling," Matt said. "I can't lie around all day. We have church on the Sabbath." David was shaking his head and was now not even hiding his grin. "I thought I could learn to ride on Thunder," Matt said. "Would that be possible?"

"Why not a more modest animal?" Will asked.

"I'll be buying Thunder once my business is complete in Richmond," Matt declared. An overwhelming feeling of satisfaction washed over him with the realization that he was taking another step toward remaining in 1762.

"Buying Thunder?" Grace said, surprised. "When did you decide this?"

"The first day I met him," Matt replied. "I don't know if you remember."

Grace looked softly at him for a moment, and then put a pensive look on her face. Matt expected her to list ten reasons why he shouldn't learn to ride on Thunder, but was pleasantly surprised when she finally spoke. "Thunder would be a good horse," she said to her father. "He's big, but gentle and easy to control."

"I still need someone to teach me," Matt said, "and the price." He glanced at Thomas, who was looking intensely at his daughter as she stared down at her food, dreading what he might say.

"Grace will teach you, Mr. Miller, should she agree," Thomas said.

"I can't teach if I can't ride," she said.

"Don't be impossible, daughter!" her father said. "I'd not expect you to teach without riding. Pledge to me that you'll take care; I can't bear to have another taken."

Tears crawled slowly down Grace's face. "I pledge this, Father."

# CHAPTER 25.

## HORSE'S ASS

He stands there. At first, the stars are visible in the horizon of the night sky and then they begin to grow into jagged balls of light. All at once they start to move directly at him, shot from an unseen cannon. He tries to dodge the speeding fireballs, but cannot prevent himself from being engulfed by burning embers that collide with his face and explode around his body, causing unbearable pain. He grows nauseous from the beating until he is holding his stomach and retching. In between coughs he looks down at his feet to see the contents of his stomach begin to take form and swirl in a vortex. The nausea is overwhelming and he tries to steady himself against the swirling. The fireballs continue to pummel his head.

Another vortex forms at his feet, and then another, and they begin spinning against each other. He can feel the shapes in each vortex vibrate with movement as they pull at his body with invisible tendrils. In a single instant his entire body shoots forward into one of the vortexes. There is an intense pain as he is torn away from the other pictures that fought for his body and he feels an overpowering sense of loss. *Was that what could have been?*

He forces himself to open his eyes and observe the vortex as he tumbles forward. Ahead of him, the shapes take on greater form and begin to move. His connection with them grows strong and there is pain as his life force is ripped from his body to feed their vibration. He reaches his hand out, but the shapes remain out of reach, hidden in the shadows. He strains his arm forward, spreading his fingers in desperation to grab them. *If I could only pull them close, I would know the future! I would know what I am to do!* There are thunderclaps, a flash, and his world saturates with light. The visions are gone.

Matt shuddered awake, disoriented, with the dream still vivid in his mind. The sun was already high in the sky and a large beam was shining through the window onto his face. *Has the rain stopped?* The barn was unrecognizable in the midmorning light, making him wonder at first whether he was still dreaming. Even the dog, still sleeping at his feet, seemed somehow out of place. Matt had grown used to him scurrying out of the barn at sunrise.

Matt's dream was starting to slip away with his other thoughts, so he made an effort to try to commit it to memory and focus on some of the moving pictures he had seen. This was the third instance where he had dreamed that he was moving purposely through a timeline. He focused on some of the moving pictures and events. Now that he knew that he had been transported into a different century, visions of the journey didn't surprise him and it made perfect sense that his subconscious mind had recorded details.

The dog opened his eyes and let out a big yawn.

"Good morning, dog," Matt said. Scout's ears perked up and he popped to his feet and headed to the door.

"I'm coming," Matt said as the dog waited impatiently. Matt sat up. Pain shot through his upper body, his head pounded hard, and the stitches in his back throbbed. Matt fought off the pain, pulled on pants, and crossed the straw-covered floor to open the door for Scout, who squeezed through and bounded out of sight. He'd be gone until it was time to bring the horses in for the night.

"You're welcome," Matt called after him. He slipped on his hiking boots and a shirt and stuck his head out the window. It was cloudy again, but the rain had stopped. He could see Joshua eating from his trough under the lean-to David had built in the nearest corral. Matt walked over the wet ground, first to visit the privy and then to see how Joshua was doing. There was a clean and fresh smell to the wet air.

Matt had just entered Joshua's coral when he heard Grace say, "How's he doing?"

"I'm hoping he looks better," Matt replied, walking towards the horse.

She walked over to the gate and let herself in behind Matt. "He does," Grace said. "I already looked."

"Really?"

"That's what I said," Grace answered.

"It's a Philadelphia thing," Matt said impatiently. Talking with her was making his head throb even more.

"What?"

"People in Philadelphia say 'really' to affirm what you said."

Grace stared with a no-nonsense look that cut through him. "Yes, it really looks better," she said. "Can we see his back now, or do we need more lessons on how people converse in the North?"

"You really don't like me," Matt said.

"I really don't know you," she replied. She looked at him, shaking her head. "Take the linen off, you'll be pleased."

"Really?" He had given up. She felt like a lost cause to him, and so he had resorted to being an ass.

"Yes, really," she said, resigned. There was some sadness in her voice.

"I'm sorry," he said immediately.

"For what?"

"Acting like a horse's ass." He said it confidently and it was obvious that it was heartfelt. His mood, his head, and the situation had combined to make him act like a malicious child.

"You're forgiven," she said. Her smile warmed to him. "Horse's ass. I like that."

Matt reached over and pulled up the linen. "We might as well leave this off until we put another coat on," he said as he rolled up the cloth. "Wow. It does look good." The swelling was beginning to disappear. Matt had no idea what the horse would look like when the infection was gone, but this was probably close. "We've got it."

"Do we put more on?"

"That's where most people make the mistake. He needs to be treated for the full ten days. If anything happens, I get delayed in town, run over by a wagon train, anything, you make sure the medicine gets put on, okay?"

"Nothing's going to happen to you in our sleepy little town, Mr. Miller," Grace said.

Matt looked at her out of the corner of his eye and pointed up to his bruised face.

"I'll carry on when you've gone," she said, laughing.

"We should wash him before we apply more ointment," Matt said. "I can do it."

"Are you familiar with washing a horse?"

"Soap, rinse and dry. How hard could it be?"

Grace put an amused smile on her face. "You've never washed an animal before, have you?"

"I had a cat growing up. But we never had to give him a bath."

She laughed like a young girl. "I suggest you never try to give a cat a bath," she said, "unless you want your arms scraped up more than they are now."

"Are you going to help me clean him or not?" Matt said.

"It's going to be your first lesson. You can't saddle a horse until it's been groomed."

"Can't the grooming wait until later?"

"Do you really think I want to groom your mount and mine?"

Jonathan interrupted them. "Good day, Mr. Miller…Good day, Grace."

"What's new, Jonathan?" Matt asked.

Jonathan thought for a moment. Matt liked this about the boy. He hadn't gotten around to adopting meaningless social banter. He really was going to tell you what was new. "Not much. Father's cleaning the horses. They got into the mud."

"Why aren't you helping?" Grace said.

"I'm *here* to help," the boy replied.

Matt looked at him out of the corner of his eye. "You're here for the improprieties, aren't you?"

"Not for improprieties," Jonathan said. "To prevent them."

"You've learned to say the word," Matt said. "Now you need to know what they are."

"Easy," he exclaimed. "It's kissing, holding hands, and all those other things."

"There's a lady present," Grace warned. She said it in a way that Matt knew she wasn't joking.

Matt motioned Jonathan into the corral. "You might as well help groom."

"We need plenty of water," Grace said, motioning to the horse.

"How much is this lesson going to cost?" Matt asked.

"One shilling," Grace replied.

"I pay one shilling to wash your horse?" Matt exclaimed.

"You're free to hire another," she said with a clever smile. "I'm sure the Paynes could instruct you."

"Funny," Matt said in a sarcastic tone.

"I thought so," she replied. "You boys ready to carry water?"

They spent the next two hours going over the subtleties of grooming. Matt tried to hurry things along by emphasizing his injuries, but Grace mostly ignored him. Joshua looked like a new horse when they were done. It was warm enough that he was dry by the time they had their supplies back in the barn, so they were able to spread more ointment on his back.

"Other horses need my attention," she said, walking away.

"When can I actually sit on a horse?" Matt said, calling after her.

"Tomorrow, provided that you return from Richmond in one piece. Will has some conceit that you'll be taking a meal at the King's Tavern." She disappeared around the corner.

"Are you all going to fight Levi again?" Jonathan asked.

"Hopefully he's in no condition," Matt said. "I don't even know if he'll be there."

"The Paynes go there every Sunday," Jonathan said.

"I don't think you have to worry. Both Nathan and Paul seem like reasonable men."

"I used to like Paul when he was going to marry Kathryn," Jonathan said. "We would have been brothers if they were married."

"Where was Levi when this was all going on?"

"He was around sometimes," Jonathan said. "He'd always try to make me play games with him and fight."

"Fight?"

"He said a boy should be tough. He didn't know how to play, though."

"What'd you mean?"

"It's like Will does with me ofttimes. You know, punching and stuff. It's fun."

"And?"

"Levi's punches hurt," Jonathan said. "He always forgot that he was playing with a boy half his size."

"There's one thing we agree on," Matt said. "Levi should learn to not punch so hard."

"I don't think anyone could teach him," Jonathan said. "Sometimes his eyes look like Shadow's."

"What? Like the horse?"

"Like he's crazy or something," Jonathan said. "One day I'll be big enough."

His fists were clenched.

# CHAPTER 26.

# ST. JOHN'S CHURCH

---

It was the Sabbath and Matt dreaded waking up. He sat up slowly in his bed, hissing in agony. It was better than yesterday, but not by much. He had taken two more ibuprofen tablets the previous night, but miracle drug that it was, it would still take more than a few days for him not to grimace in pain at every movement. The dog turned his eyes to watch.

"Good morning, dog," Matt said. Scout picked his head up and looked at the closed door. Matt shimmied to him and scratched his head. "I ache," he said. "You've my permission to bite that bastard next time you see him." The dog moved his head around for Matt to scratch the other side. It occurred to Matt that this was the first time he had actually touched the dog.

"I guess I should start getting ready for church," Matt said. "I'm dressing as a colonial gentleman today, new shoes and all."

The dog tilted his head in that way he did, like he understood.

"Don't look at me like that," Matt said, smiling. "I can pull it off."

Matt hopped onto the floor, stepped into his hiking boots, and made his way to the privy. The sun was starting to be visible on the horizon. The privy was cold from the night air and he was wearing only a cotton tee shirt, so it was impossible to get comfortable. He got up as soon as he was finished and went back to the barn to wash. He filled his basins at the well, brought them back to the barn in two trips, and shut the door. He stripped down and washed his entire body with ice-cold water, used shampoo from his pack to wash his hair, and then shaved.

It took him the greater part of an hour to complete all of this. He felt his face, and it seemed smooth. With no mirror, there was no way to tell if his sideburns were even. He finished up by combing his hair, which was starting to get long. A haircut would be another thing to get on his next trip to Richmond, unless this was something one of the Taylor women could do. Giving a reasonable haircut seemed like a skill a colonial mother would have. He dried his body and pulled on a pair of pants. There was a knock on the door shortly after.

"Mr. Miller, are you awake?"

"I'm awake," Matt called out. "Thanks."

"Can we come in?"

"Open the door," Matt called. The door slid open and the two youngest Taylor boys walked in.

"Hi, Mr. Miller," they said in unison.

"How'd you sleep?" Matt replied.

"Very well!" Jonathan said. "Hi, Scout!" He walked over to pet the dog.

"What about you, Jeb?" Matt asked the older boy, who was looking unsure of himself.

"I slept well, I guess."

"Is it Sara Greene?" Matt asked.

"I can't stop thinking about her!" Jeb exclaimed. "Something is wrong with me."

"That's what they do," Matt replied, laughing. "They drive you mad."

"Even when you're as old as you?" Jeb asked. He wasn't joking.

"Even as old as me," Matt said with some chagrin. "I'm only twenty-six."

Jonathan gave Matt a big smile and Matt returned it with a fake scowl. "Don't even say it," Matt said.

"Jeb can't stop thinking of Sara Greene," Jonathan said, "and Mr. Miller can't stop thinking of Grace."

"Oh, everyone knows that," Jeb replied matter-of-factly.

"What do you mean, everyone knows?" Matt said, surprised.

"Do you have trouble sleeping?" Jeb asked.

Grace had continually occupied Matt's thoughts during these last days and there were times when he would lie in bed trying to think of her entirely, but usually he was so tired at the end of the day that he had trouble keeping his eyes open. *Maybe I am getting old.* "I have no comment," he said, chuckling. "Anyway, I don't want to talk about Grace."

"Did you two quarrel?" Jeb asked.

"No, we didn't quarrel," Matt replied, incredulous. "We don't even know each other."

"You should make a better acquaintance," Jeb said. "That's why I speak to Sara."

"Did you two come to discuss my love life, or was there something else?"

"You love Grace?" Jonathan said.

"No, I don't love Grace," Matt replied quickly.

"You admire her, though, right?"

"Yes!" Matt said, slipping.

"When will you tell her?"

"I'm too old to tell a girl I *admire* her."

"Should I tell Sara Greene that I think of her constantly?" Jeb asked.

"No! You gotta be cool."

"How does being cold help?" Jeb asked.

"It's something we say in Philadelphia," Matt said. "Being cool around a girl means you're all calm and confident. You act like you aren't impressed she's talking to you."

"I can do that," Jeb replied.

"Don't be too cool, though. You must still act interested."

"Being cool sounds complicated," Jeb said.

"Sara Greene doesn't always ignore you," Jonathan observed. "She must admire you a little."

"You're a good-looking young man from a good family," Matt said. "What's not to admire?"

"Jeb wants to use the toothbrush," Jonathan said. "The ladies love a man—"

"Don't repeat that," Matt said. The boy smiled. Matt handed his toothbrush over to the boys. He cleaned up his toiletries while they finished brushing.

"Did you use the toothbrush, Mr. Miller?" Jonathan asked. "It may work on Grace."

"Don't push it, kid," Matt said, giving him a stern smile. "Before you go, I need two things. First, look to see if my back is bleeding, and then tell me if my sideburns are straight." They took the next couple of minutes trimming Matt's sideburns and putting a bandage on his back.

The dog watched the boys go and Matt shut the barn door behind them. "Interesting characters. Let's get these new clothes on."

He spent more time than he expected figuring out how to get dressed, as the dog looked on with seemingly profound interest. Matt dressed in white stockings, dark grey breeches, a white shirt, and then the new black shoes. He tied the cravat as well as he could and then finished it off with a dark grey waistcoat that matched the breeches and the tricorner hat Will gave him. He didn't have a mirror, but he imagined that he looked reasonably colonial.

"Ready to go?" he said to the dog. Scout hopped off the bed and followed him out the door. He stayed by Matt's side all the way to the house, then took his position on the porch. Matt knocked on the door and it opened soon after.

"Good morning, Mr. Miller," Grace said. She wore a blue dress with white trim and matching blue ribbons in her hair. The ribbons highlighted her blond hair and made her eyes look bluer.

*A drop-dead beauty!* Matt smiled. "Where is everyone?"

"Late start," she replied. "Father and Uncle talked into the night."

"Anything important?"

"You'll dine with them today?"

"Wouldn't miss it."

"It doesn't vex you that Levi will be there?" she asked.

"Sure, it *vexes* me," Matt said. The word felt weird coming out of his mouth.

"He's expressed his desire to kill you," Grace said. There was concern in her voice that pulled him in.

"He'll think twice," Matt replied. "He knows I won't back down."

"Every quarrel need not always end in you striking someone," she said.

"Should I run away like a coward, then?"

Grace looked at him silently, impossible to read. "Shall I fix your cravat?" she finally asked.

"I didn't have a mirror. I must have tied it five times trying to get it right."

"I tie them for my brothers," she said. "Come hither."

Matt walked forward, and Grace reached up to fix his necktie. She was agonizingly close as she set the length and began tying the silk around his neck. He drank her in as he cataloged every detail of her face, from her blue eyes to her nose, to her full red lips and the smooth white neck that peeked out from under blond hair. Every casual touch of her fingers shocked him into some higher form of consciousness and it took all his control not to pull her close and kiss her passionately. The scientist in him wondered what it was about her proximity that affected him so profoundly. Did she give off an electric field or pheromones that made his knees weak?

Grace finally stepped back and looked at her handiwork. "Perfect."

He had to consciously wake himself from the spell she had cast. "Thanks, Grace."

"Maybe Graine will be there," Grace said, teasing.

"Henry assured me that I'll be very fashionable in these new clothes," Matt replied, "and that I would impress the ladies."

Grace looked him up and down. "You may capture Graine's attention," she replied. "You're the type of man she fancies."

"What kind of man is that?" Matt asked. He was intrigued to hear her assessment.

"Bookish, but handsome," she replied.

"Bookish? That's a horrible thing to say about a man."

"'Tis a compliment." She thought for a moment and said, "Men of letters are usually wealthy. We'll not tell her you work as a farmhand...or the drunkard part."

"Funny girl," Matt replied.

"I should not have believed it. I'm truly interested in whether you can charm Graine. It would speak much to the skill of Henry Duncan."

"Funny...again," Matt said sarcastically. "Some women find me charming even without new clothes."

"There's someone for everyone, I imagine," she said.

*You stuck-up little...*

"Ready for church?" Mary said, walking down the steps. She stopped completely when she saw Matt and said, "Henry Duncan has outdone himself. You look very handsome."

"Thank you," Matt replied. "He had a lot to say about dressing as a Virginia gentleman."

"He's a charming man," Mary said.

"Very boisterous and animated," Matt replied.

"He's the biggest fop in the city," Will exclaimed, coming down the steps.

"You would never speak thus in his presence," Grace said.

"Do you believe your brother mad, sister?" Will replied.

"There have been times when I have questioned your judgment," she replied, smiling.

Matt was puzzled. "What would happen if you said it to Henry?"

"He'd thank you if he was in a good humor," Will explained, "or slice you into tiny pieces if he was otherwise."

"Are we talking about the same Henry Duncan who talks endlessly about London fashion?" Matt asked.

"He's a master swordsman," Will said.

"I would not have believed he was a dangerous man," Matt replied.

"He's the friendliest man in Richmond," Mary said.

"I jest. We all admire Henry," Will said. "We wouldn't have sent you there otherwise."

"It was a good call," Matt said as he fumbled with his hat.

Will motioned for Matt to hand it over and then demonstrated how it should be carried under his arm. He returned the hat to Matt. "That's how a Richmond gentleman carries his hat." He watched Matt practice.

"Not exactly," Grace said. She stepped to Matt and adjusted the hat lower in his arm. "It makes you look more relaxed and confident if it's not so high." The hat slipped from Matt's arm onto the floor. She said, "You should master that for Graine."

"Dear sister," Will said. "Are you playing matchmaker?"

"I wish to learn if clothes truly make the man," Grace said, smiling. Matt gave her a good-natured frown.

"Everyone ready?" Thomas said as he walked into the room. "Mr. Miller, you look a man of parts. The cravat is perfect."

"I had help," Matt said, nodding toward Grace.

"She ties everyone's silk," Will said, laughing. "'Tis why we keep her here, unmarried."

"That's quite enough," his mother scolded.

"She knows I jest," Will replied. "If she weren't so hard to please."

"You've had no greater success," Grace said. "Twenty-five and unmarried!"

"I've not found a maiden who appreciates my humor."

"Why have you children waited so long to be betrothed?" Mary asked.

"I have not seen anyone worthy of either," Thomas piped up.

Mary lightly slapped him on the shoulder. "With such conceit," she exclaimed, "I shall never have grandchildren for my knee."

"I dare say neither of these children will have a problem," Thomas said. He turned to Matt and looked him up and down. "I hope Henry is there to see the results of his handiwork."

\*\*\*\*\*\*\*\*\*

They drove to church in two separate wagons. Matt's was pulled by two dark brown mares and black stallions pulled the wagon carrying Thomas and the women. The mares weren't twins, but it was obvious that some thought had gone into picking closely matching horses.

"Are these four horses for sale?" Matt asked.

"Will you purchase them as well?" Will said, smiling.

"It's a serious question. I'm interested in how much horses cost."

"We desire forty, perhaps forty-five pounds each," he said. "Would you sell them, then?"

"I could use the commission after all the money I spent," Matt answered jokingly.

"You picked the finest cloth," Will said, looking at Matt's jacket. "Graine *may* be charmed."

"So I've heard."

"I regret my jesting," Will said. "She is a woman of excellent wit and I have thought she was very attractive."

"Maybe you should marry her."

"Perhaps."

# CHAPTER 27.

# GRAINE

St. John's Church was a beautiful white building surrounded by a matching white picket fence. Will drove the wagon through the gate and into the parking area. He hopped onto the ground and motioned for Matt and Jonathan to follow. Will handed the reins to one of the black men who was attending the parking area, turned around and led them to where the Taylor family stood waiting.

"Two Sundays in a row, Mr. Miller," Grace said. "One more and you may be required to purchase a seat."

"Have your fun, Grace," Matt said. "I was looking forward to coming."

"I thought you didn't attend a church," Grace replied.

"I don't," Matt replied. "That doesn't mean I don't wish I did."

"Why not attend?" she asked. There was no judgment in her voice.

"If you don't grow up in church," Matt explained, "and none of the people around you belong, then you don't even think about it."

"You'll have a finer wife should you attend church," Grace declared.

"Who says I'm looking for a wife?" Matt gave her a sly grin.

"Aren't all gentlemen looking for the perfect lady?"

"If you see her," Matt said, "let me know." He gave her his best poker face.

Mary led them through the church doors and then to the front to their regular box of seats. Matt followed, thinking about how handsome he must look in his new clothes. He noticed people, especially women, staring at him.

"Mr. Miller!" Jonathan whispered, tugging on his arm.

"What?"

Jonathan pointed at his face.

Matt thought he must be commenting on his bruises. "They don't look that bad."

The boy whispered, "No!"

"What?" Matt asked again. They were almost ready to sit down. Matt was getting irritated.

Jonathan pointed again and this time mouthed the words slowly as he said them. "Your hat!"

"My hat? What's wrong with my—"

"Take it off," the boy said louder.

"Oh," Matt said in surprise. He reached up and snatched the hat from his head. "Thanks," he whispered. He made a mental note to buy the kid some candy.

They shuffled into the box in single file and sat down. Grace handed Matt some crudely printed booklets so he could pass them to her family. Eventually, the preacher stepped to the wooden lectern and shuffled his notes. "I see some new faces in our church today," he said, looking directly at Matt.

Matt smiled back with his best *Hello*.

"That's good because it's not my desire this day to speak to the converted," the preacher said. There was a dramatic pause and then he went on. "Our Lord didn't spend his time with the religious men of his time; he despised them. No! He spent his time with the prostitutes, the tax collectors and...the drunkards!" Grace turned her head towards Matt, and then looked back at the preacher. "You should see the dismay on Miss Taylor's face," the preacher said to the congregation, laughing. Chuckles sounded through the building. "She suspects her young man has been already convicted of a long list of crimes. Poor lad, you'll learn that there's naught worse than a lady's suspicion." There was good-humored laughter all around. Matt could hear shuffling as people strained to see the young man in question.

"Fear not, Miss Taylor," he said, looking now at Matt. "I know him little more than Adam, but it's plain to see that he's a sinner and he'll sin again." The preacher turned back to Grace, and it was obvious from his knowing smile that her expression hadn't changed. Matt's body felt paralyzed and he had to focus to be able to move his head enough to see Grace. He saw her face and immediately turned back to look at the preacher. The preacher took another dramatic pause, then gazed out at the congregation and said, "There's naught that is special about Miss Taylor's young man, because we are all sinners in the eyes of the Lord. It's only through Jesus that we are saved."

He then went on to shout fire and brimstone for the next twenty minutes, but never mentioned Grace or her "young man" in his sermon again. Matt had never seen such a riveting speaker. Now and then, the preacher would take a breath, and it was during these times that he'd look down at them both with a knowing grin. Matt tried his best to smile and he chanced a few quick looks over at Grace to try to assess her state of mind, but she stared straight ahead.

When church was over, they shuffled out for fellowship in the churchyard under the shade trees. Matt found himself alone as he had been the previous week as the Taylor family made their rounds. Henry Duncan stepped to him. He was silent at first as he judged the fit of Matt's clothes. His satisfied grin disappeared when he finally focused on Matt's face. "Oh my!" he said.

"I got in a fight after I left your store," Matt explained.

"You fought Levi Payne? They claim you kicked him. Is this some particular fighting style?"

"Something I learned in China," Matt replied.

"Would you mind teaching me? I'm a student of the fighting arts."

"I heard you were a swordsman," Matt said, nodding. "It's called tae kwon do. It's a self-defense technique."

"Many of the Oriental styles I've studied are defensive," Henry replied. "When can you visit?"

"Next weekend, maybe," Matt said. "I have another week committed to the Taylors."

Henry looked at him, laughing. "I imagine you're Miss Grace's young man."

"I don't think she's very pleased with all that," Matt replied. "It's a mystery how the preacher made the connection."

"He's called a rector," Henry said. "The Reverend Michael is a good judge of people. There may be more in her manner than you see." Henry thought for a moment. "Speaking of," he said. "Have you met Graine Martin?"

Henry pointed over Matt's shoulder. "She's a formidable young lady."

Matt turned to where he was pointing, and as if on cue, Graine looked up to see them staring. Henry waved to her, and Matt smiled. Graine put her hand on the arm of the woman she was talking to, said something, and then walked over.

"Good morning, Mr. Duncan," she said as she offered her hand. He leaned down and lightly kissed it. She turned her attention to Matt and extended her hand. "And 'twas Mr. Miller." She presented herself with a confidence and poise that Matt hadn't appreciated during their first meeting. He accepted her hand in his.

"You have a beautiful name," Matt said. "I might like to use it someday for one of my own daughters."

"You might want to ponder that," Henry advised.

"Whatever could you mean?" Graine exclaimed. "Mr. Miller has paid me the most excellent of compliments."

"Graine Miller," Henry said, laughing. "Unfortunate for a young lady unless she plans to enter the flour trade."

"She'd marry, and it wouldn't be an issue," Matt said.

"This much is true," Henry replied. He motioned like he was leaving. "I realize that you young people have many such possibilities to discuss." He turned to Matt. "I look forward to a demonstration of your fighting style." They both watched him walk into the crowd.

"Why did he speak of fighting?" Graine asked. She was looking suspiciously at the bruises on his face.

"I learned how to defend myself in China," Matt said.

"Why did you travel to China?"

"To learn apothecary," Matt replied. "I'll be starting a shop when I return to Philadelphia."

"Quite exciting!" she exclaimed. "I have been with my father ofttimes when he's begun new enterprises. Much can occur...'tis exhilarating."

"You sound like you've a lot of experience," Matt said.

"My father has multiple ventures," Graine explained. "I maintain his accounts."

"You're not what I expected," Matt said.

"Mr. Miller, I hardly know you," she said, surprised. "How could you expect anything?"

"Some men in town have mentioned you."

She was shocked. "And what do these fellows say?"

"They say you're beautiful, but that you spend your father's money like there's no end."

"I take pleasure in saving my father's money. What's the proof they have for this?"

"Something about your expensive clothes," Matt said. "You do dress well."

Graine smiled at the compliment. "My father runs an import business. Most of the pieces we keep are either free or discounted. It's a benefit of the shipping trade."

"That kind of explains—"

"Does William Taylor believe this of me?" she asked.

"I wouldn't break his confidence," Matt said, "but he has nothing but good to say about you." He saw her relax.

"You're the young man that the Reverend Michael spoke of today?"

"Grace wasn't too happy with all that," Matt repeated.

"You don't seem vexed."

"I don't feel one way or another," Matt replied. "Do you know Grace very well?"

"Are you trying to win her heart?" She gave a knowing smile.

"A matter for another day," Matt replied. "Seems like you two would be friends."

"We were," she replied. "Now we only say good day."

"What changed?"

"Mr. Miller," she said, irritated. "I don't believe the trivialities of my life would be of any interest."

"What changed?" Matt repeated.

"I told Grace that ladies shouldn't be on horseback," she said, resigned. "It hasn't been the same between us since."

"What possessed you to say something like that?"

"I was upset over Kathryn. She shouldn't have been on that horse."

"Didn't she know how to ride?"

"Yes, as I'm sure you're already aware," Graine replied. "That's not the issue."

"What's the issue?"

"You're a very vexing man," Graine said.

"Not with everyone. I think it's mostly beautiful women and town bullies."

She answered him with a look that said she thought he might be insane.

"I've overstepped my bounds," Matt proclaimed. "Accept my apologies." He gave a slight bow and turned to leave.

Graine put her hand out to touch his arm. "Please don't go, Mr. Miller," she said. "I said those things to Grace because I believe them. No matter how skilled, neither had the strength to handle a horse."

"Many women ride horses where I come from," Matt explained. "They don't get hurt any more than men."

"Ladies shouldn't wear breeches and ride horses!" she exclaimed.

"Nor should they manage a business."

"My involvement in my father's enterprise is different!" she exclaimed. "It's our trade."

"And so are horses for Grace Miller."

Graine became quiet and only stared back at him, and Matt was very sure that he had convinced her. "Don't you have anything to say?" he finally asked.

"You called her Grace *Miller*," she said.

"I did not."

"Will she wear breeches and ride when you're her husband?"

"Yes," he said without thinking.

"You're hopelessly in love."

"This isn't about me," Matt replied.

"Then who, Mr. Miller?" She smiled. "I've quite forgotten."

"Apologize to Grace for your comments."

"Fine," she replied.

"Simple as that?"

"If her future husband won't mind her wearing breeches and riding, why should I?"

"Funny," Matt said sarcastically.

"Mr. Miller," she said, "before you seek out Grace, give me the opportunity to engage her. Also, please don't whisk Will away before I've had a chance to let him say good day."

"Graine," Matt said, kissing her hand, "one other thing."

"What's that, Mr. Miller?"

"Do you want to buy some horses?"

# CHAPTER 28.

## SIT-DOWN

---

Much to Matt's consternation, by the time he was able to talk to Grace, they were about to leave.

"Will has been searching for you," she said. "They are ready to retire to the King's Tavern."

"What're you doing while we're away?" Matt asked.

"A picnic with the Martins."

"Graine Martin?" he asked, surprised that things had moved so quickly.

"We finally moved beyond an animosity that had developed between us," Grace replied.

"Are you going into the country for the picnic?"

"Why would we do that? The picnic's at the Martins' house. Graine mentioned that her father had recently imported some new dresses from London. She'll show them to Mother and me."

"So now you're in the market for expensive London fashion?"

"As I know you are aware, much of her clothing costs significantly less than I believed. She offered us clothing at their cost."

"I begged her to help you," Matt said. "She does dress beautifully."

"So you're helping me?"

"I live to serve," he said, bowing.

"There you are," Will said, walking to them. "You sold two more horses?"

"Sold and bartered," Matt said. "I'm exceedingly proud of the barter part, for obvious reasons." He said this while putting a critical expression on his face and looking at Grace's dress.

"That's true," Will said. "I'd start practicing, sister." He went silent.

"I know you prepare me for some jest," she replied, "but you'll not rest until it's out of your mouth." She put on a falsetto voice. "My word, practicing for what, dear brother?"

"Fanning," replied Will. "I'm simply glowing." He flapped his hand at his face, laughing hysterically.

"Enjoy your diversion," she said, "but it won't keep me from visiting with Graine. It's been overlong since we've spoken."

"Why is that?" Will asked.

"As Mr. Miller is fond of saying, 'tis a long story," she said, looking past Will. "I see Mother. Have a capital time at the tavern. No fighting." She turned and left with a sharp swish of her dress. Matt could only watch and smile. *Stunning woman.*

"You've done it again," Will proclaimed.

"Done what?"

"Gone and made my sister happy."

"It's hard to tell with her sometimes," Matt replied. "But if she is happy, I'm glad. She did seem pleased with getting new dresses."

"She's pleased to renew her friendship with Graine," Will said.

"Graine should be an exciting addition to the Taylor family."

"You intended to say Miller family, of course," Will replied.

"No, Taylor family," Matt said. "Could there be a prettier woman in all of Richmond?"

"Probably not," Will said, resigned.

"You don't admire her?"

"I do, but I'm not ready to marry."

"Nor is she, I imagine," Matt said. "She had that dreamy look in her eye when your name came up, though."

Matt saw Will become distracted and scan the crowd. "As much as I endeavor to speak of beautiful ladies," Will said, "we must focus on the charge at hand." He motioned for Matt to follow him toward Thomas, who was talking to another man in the churchyard. He shook hands with the man and then walked to meet them. "You boys prepared for our interview?" he called.

"I imagine," Will said, reluctantly.

"Would you rather stay and pick out dresses?" Thomas asked.

"Buy a nice satin one for Levi," Matt said, "and suggest he try it on in the tavern."

"There is some hidden lunacy in you," Will said. "Best tell Father your history with Levi." Matt took the next few moments to describe what had transpired the first time he met Levi in the King's Tavern.

"You both antagonized him," Thomas said after Matt finished. "You declared you labored for free and were bringing others for what?"

"Fresh air," Will said. "'Twas not entirely Mr. Miller's fault."

"Mr. Miller surely played his part," Thomas replied.

"I'm sorry, sir," Matt replied. He *was* sorry.

"What's done is done," Thomas replied. "Either way, Levi is a damn fool. Keep clear."

"I'll do my best," Matt said, "but I can't promise anything."

"You must," Thomas said. "Fighting in the tavern is unacceptable. If you should desire later to have yourself killed, then that's your choice. This day, though, naught comes from you. Do you agree to these terms?"

"I promise," Matt said. "You're talking to me like I make my living fighting!"

"I'm comforted knowing you'll be there if there's trouble," Will piped up. "You can kick him."

"You heard me," his father said. "No fighting! Neither of you will say anything beyond *good day.*"

"We aren't allowed to talk?" Will asked, surprised.

"Naught," his father repeated. "No matter what Levi does, you're to remain silent. Let the Paynes be impulsive this day, not us."

"What if he insults Mother?" Will said. "Can we fight him?"

"He'd not insult—" the father said, then stopped himself to laugh. "If he does, you both have my permission to put up the gauntlet, else not a sound from either of you."

"What if he insults Scout?" Will said. "Can we insult one of his animals, a cow or something?"

"This is no jest," Thomas said. "There are serious consequences. The thought of Levi threatening our men or our business has me very concerned."

"Neither of us will say a word," Will said as he looked over at Matt for confirmation. Matt nodded.

"We're taking the stallions," Thomas said. "Graine wants to show her father the mares. She's a wonderful young lady with a good head on her shoulders." He focused on his son. "Her family is as solid as they come."

"Are you and Mr. Miller in some collusion?" Will asked. Thomas looked at his son, puzzled, and then he turned his gaze to Matt.

"I told him the same thing," Matt declared. "I didn't know the family part; I assumed."

"Paul Payne might be a match for her," Thomas said.

"She'd never be interested in Paul," Will replied.

"Don't be so sure," Thomas said.

"Let me think on it," Will replied.

"Think on it as you desire," his father said. "Realize, though, that there are more than a few eligible men in Richmond. I'll grant that none has so grand a title as Horse Prince."

"I told you, that's only a jest between me and my fellows," Will said. "It's Prince of Horses, anyway."

Thomas rolled his eyes. "Let's repair to the tavern," he said. "I shouldn't want to miss them."

"Such a tragedy," Will said sarcastically.

"It gives me no more quiet than you," his father said. "We should discover the gravity of our situation. I suspect 'tis much worse than we know."

"Why do you say?" Will replied. "'Tis only made worse by some aversion Levi has to Mr. Miller."

"Something's changed," his father said. "Levi's conduct is more than is common."

"Maybe we can eat before we fight to the death," Will proclaimed.

"I'm all for that," Matt said. "I don't want to go through the pearly gates on an empty stomach. I'd be all cranky, and that's no way to start off a stay in heaven."

Both Taylors turned to stare at him as they walked to the carriage, shaking their heads. Matt smiled back.

**\*\*\*\*\*\*\*\*\***

They arrived at King's Tavern more quickly than Matt had wanted. Thomas guided the carriage to the stable master. Inside, the tavern was buzzing with activity, much like the previous week. A bar hostess walked to greet them. "Anne, dear, could we get something in the back?" Thomas said. "We desire a table to seat six fellows for an interview." He handed her a coin. She motioned towards two men at the bar, who went to prepare the table and then guided them to the dining room. Thomas turned to Will and Matt and said, "Save a chair."

The Paynes were at their regular table. Matt met their gaze as he walked by, and in that moment it felt like the true nature of the Paynes became clear. He imagined these men like a pack of wolves trying to separate him from the herd. "Don't stare," Thomas said, seeing him looking back. "Go to the table and let me take care of this."

"Zounds!" Will whispered loudly. "You broke his nose." Levi's nose and face were black and blue. It made Matt strangely satisfied; he'd had some difficulty remembering whether he'd hit Levi at all. Matt and Will split off from the older man and headed to the table, but Will watched his father protectively the whole time they were walking.

Nathan Payne rose as Thomas approached their table and shook his hand, and they stood talking. Thomas often glanced at his son and Matt during the conversation, and then would look back to Nathan. They were apparently trying to agree on something before coming over. Then, both Payne sons stood up and they walked over as one. Matt found that his fists were clenched and his legs were locked, so he took a few deep breaths and forced his muscles to relax as his tae kwon do instructor had taught.

He and Will remained standing as the men approached the table. Thomas joined them and Nathan took his place with his sons. The six men faced each other and didn't attempt to shake hands. It was all very bizarre, and if the situation weren't so serious, Matt would've broken out into hysterical laughter. The situation became no less strange when he found himself standing in front of the man who wanted to kill him. Matt knew that the best way to unnerve an opponent was to look directly into his eyes and not blink, and so he returned Levi's glare with unblinking eyes. He wasn't disappointed when Levi was the first to talk.

"What are you staring at?" Levi said.

"Levi!" Nathan said. "You'll not fight in here."

"He won't look at me, then," Levi replied, "if he wants to keep his teeth." Matt slowly turned his back foot into a fighting stance. "I'll kill you, you son of a bitch," Levi said.

Nathan put his hand up. "Stop. It will serve neither family."

"Mr. Miller!" Thomas said. "Enough!"

Matt shuddered. He backed away from the table and nodded to Thomas. He focused on a picture on the wall over Levi's shoulder but kept Levi in the corner of his eye. He felt like he had been in a trance and time had slowed. He'd been able to see the other man's eyes blink through their entire motion.

Thomas gestured to the table. "Please," he said. "Sit." He sat first, and the rest followed. "I have instructed my boys not to speak unless asked. The posturing of young men should not spoil what I hope will be a beneficial transaction for our families."

"He's not your relations," Nathan said, glaring at Matt. "He's not even a Virginian."

"I decide who represents my family," returned Thomas.

"He provoked my son and caused a fight," replied Nathan.

"Levi fights when the wind blows," Thomas said.

"We're here to be insulted?" Nathan replied.

"We're here because a chasm has formed between our families," Thomas said.

"That your people have caused," Nathan retorted.

"We concern ourselves with our own affairs," Thomas replied. "Explain how we caused this?"

"My son is vexed," Nathan said, "and I share his feelings."

"And this justifies attacking members of my farm and family?" Thomas replied.

Nathan waved his hand to quiet him. Thomas was visibly irritated by Nathan's dismissal. "I stood by these last few years and did nothing as my frustration with you and your relations grew," Nathan said. "I was willing to ignore many things because of my Paul's love for Kathryn."

"Nathan, I don't see—"

"You Taylors speak endlessly, but now I'll have my say," Nathan said, interrupting. "I stood by whilst you purchased the finest stock in Virginia and made alliances with all the best breeders. You've attempted to shut us out of the trade."

"You've as many opportunities to buy horses as we," Thomas said.

"How many times have I tried to hire your black horse for stud?" Nathan asked. "And yet somehow, the cost or time is never right."

"I have kept Shadow's line confined to my farm," Thomas said. "I'll make no excuses for this."

"No more!" declared Nathan.

"What do you mean, *no more?*" Thomas replied. "We do as we will under the king's laws."

"We'll no longer stand by as you purchase the finest horses in the colonies and leave us with pack animals," Nathan declared.

"How would you deter us from conducting our affairs?" Thomas asked.

"Wherever there's a horse for sale," Nathan said, "we'll offer more. I have fellows too and they'll know that when the Taylors are interested in a horse, I'll double the price."

"Double?" Thomas exclaimed. "You would break your back to act on this vendetta?"

"We've amassed a sizable fortune selling our inferior animals," Nathan said. He emphasized the word *inferior* while looking at Will. Thomas turned to his son with a probing stare.

"I'd never describe another man's horses thus!" Will cried.

"You have your fame," Nathan replied. "You declare your horses are the best in Virginia and that no others can compete. People believe Payne horses are inferior. As I said, no more!"

"This crusade will lead you to ruin," Thomas said.

"Even when you're weak, you speak as if you have the upper hand," Nathan said. "This is merely business. I'll see that you have no access to fine stock."

"It's a fool's errand," Thomas declared.

"I've already started," Nathan quipped. "That champion stallion at the Browne farm?" He paused for a reaction, but Thomas remained stoic. "I told him I'd double your price."

"Go buy him," Thomas replied. "Virginia horsemen will be pleased with your extravagance."

Nathan sneered. "I grow weary of the Taylors and how they are a pillar of Richmond. I know the real Thomas Taylor."

"Men change," Thomas said.

"You, your sons, and your daughter all think you're better than everyone."

"You'll not speak of my daughter," Thomas warned.

"How upstanding will the Taylors be when their money and power are gone?"

"You're committed, then, to this war between us?" Thomas's tone was both disappointed and tired.

"We'll be using all our means to drive you into the ground," Nathan said calmly.

"If that's your pledge," Thomas replied, "we're done here." Thomas waved to Matt and Will that it was time to leave and everyone stood. Thomas focused his stare on Nathan. "One more thing. If you or one of your sons threatens a member of my family again, I'll come and put a ball in your head."

# CHAPTER 29.

# STRATEGIES AND PLANS

They walked out of the tavern in silence. Thomas reached into his pocket for a coin and gave it to the stable master. He and his son climbed into the front and Matt took the seat directly behind them, leaning forward to hear their conversation.

"That could have gone better," Thomas said. "It's always two steps forward and one step back when following the path of the Lord."

"I swear to you," Will said, "I have never slandered Nathan's farm."

"Don't apologize for selling horses," Thomas replied. "It's easy to see that Nathan's horrible at breeding."

"How did he learn of the Browne stallion?" Will asked.

Thomas shrugged. "Of everything, this disappointed me the most. I've already been planning the foals from that one."

Will glanced over his shoulder at Matt. "You started this," he said. "Have you no thoughts?"

"I'm an innocent victim," Matt proclaimed. "The war's been building for a long time."

"You're not an innocent victim," Thomas said. "You have some hidden lunacy in your desire to tease Levi at every interview."

"I should have brought the dress," Matt declared. He had a painful smile on his face knowing that the man was partly right. He regretted his inability to ignore Levi.

"The dress?" Will asked.

"The satin one for Levi," his father chimed in.

"Levi was upset because I came empty-handed," Matt said.

"Next time," Will replied, laughing, "bring the dress."

"You were staring him down," Thomas said. "I've never seen Levi so irked."

"It's true," Matt said, "about his eyes."

"What?" Will asked.

"Something Jonathan said," Matt replied. "He has eyes like Shadow's."

"The horse?" Thomas asked.

"Wild, as if something's missing," Matt replied. "I didn't mean to insult Shadow."

"Probably accurate in both regards," Thomas said.

"We weren't able to eat" said the older man, changing the subject. "The women will expect us to have eaten."

"I know the perfect place," his son replied.

*********

Soon after, they were eating in the Gold Lion Inn. "How shall we respond?" Will asked his father.

"The Lord is telling us to dig up our talents," Thomas replied.

"Good stock may become hard to find with Nathan scattering gold everywhere," Will said.

"Nathan couldn't recognize a quality animal if it bit him on his ass," Thomas replied. "I'll enjoy watching him pay double for every horse that strikes our fancy."

"It's a shame that he bought the Browne stallion," Will said.

"I'll contact Douglas to see if the sale took place," Thomas said. "Something of Nathan's story smelled of fancy. Would he pay a hundred forty pounds? I might be prepared to spend eighty."

"Eighty pounds!" his son exclaimed. "That truly is mad money!"

"I'm not mad," his father said matter-of-factly.

"An expression Mr. Miller uses," Will explained. "Why would you think Nathan was lying?"

"His manner," Thomas said. "He was hoping for a reaction." Thomas was quiet, clearly considering something. "Maybe I'd pay ninety pounds," he said. "I don't like this 'mad money' expression you boys are using, though."

"He's worth ninety pounds?" Matt asked.

"I'd not usually pay that much for any horse," Thomas said, "but there's none better than Douglas Browne when it comes to breeding."

"What's so great about this horse?" Matt asked.

"He's Shadow without the empty eyes," Will replied.

**\*\*\*\*\*\*\*\*\***

It took about twenty minutes to make their way to the Martin house. The Martins lived in an opulent home with a long drive, right outside of Richmond. Matt could see people gathered off to the side of the house and a smoking fire where slaves cooked food. Thomas pulled the wagon up to the front of the house, where they were met by a well-dressed black man. Thomas handed him the reins and the man climbed up into the wagon to move it around the side of the house, out of sight. Thomas gazed at the enormous estate and said with a sarcastic smile, "It's a humble shack, but we call it home." Matt and Will laughed with him as they walked toward the picnic area.

Robert Martin came to great them. "Good day, Thomas," he said, reaching out to shake his hand. "Sorry I missed you at church today. There was some crisis at the dock."

"Business calls," Thomas said.

"It occurs too often these days," Robert said. "I'll increase my tithe this week to prevent the reverend from making me the subject of his next sermon."

Will reached his hand out. "Good afternoon, Mr. Martin."

"Good afternoon, Will," Robert replied. "It has been overlong since we've seen you."

"Samuel keeps him busy," his father chimed in. "We rarely see Will."

"Learn what you can," Robert said. "There's no better man with money than Samuel."

Thomas spoke up as he motioned to Matt. "Robert, please meet a family fellow," he said. "Matthew Miller."

"Pleased to meet you, Mr. Miller," Robert said as he reached his hand out.

"It's my pleasure," Matt replied. Robert Martin's handshake was firm and warm. He seemed altogether comfortable with himself and his place in the world; it was obvious where his eldest daughter had learned her poise.

"We bought some of your horses," Robert said, looking back at Thomas. "What fine animals! We are stealing them from you at that price."

"They are some of our best," Thomas said. "All should feel that they stole their horses from us."

"This may be the case," Robert said, "but I have instructed Graine to make up for this with fine dresses. It's been difficult to convince my eldest that both parties should profit in any venture."

"They walk on air with new dresses," Thomas said. "Are they still choosing?"

"Grace is in the stables with my Graine and those two new horses," Robert said. "The ladies mentioned something about learning to make them shine." He shook his head. "Graine spends too little time on activities that endear her to other ladies."

"Sounds like Grace!" Will said. Matt saw Thomas give his son a disapproving look.

"You jest, of course. Your sister is as feminine as anyone I've seen," Robert said. "If I had a son, I'd be lobbying this very moment for an alliance."

"Thank you, Robert," Thomas said. "Sometimes my son speaks from passion, and it would not do for his sister to hear." He gave his son another dirty look.

"Your secret is safe with me, lad," Robert said to Will. "I should wish to speak more to your father. Graine instructed me to send you both to help with the horses when you arrived."

*********

As Matt and Will rounded the corner to the stables, they saw the girls working with the two mares. Grace was showing Graine how to groom them with a fine-bristled brush to bring out their luster. "Learning to brush your horses?" Matt called as they approached.

"I gave Mr. Miller the same lesson yesterday," Grace replied, smiling, "but I charged him a shilling."

"Whatever for?" Graine said.

"I'm teaching him to ride," Grace explained.

"Would you teach me?"

"What?" Grace didn't attempt to hide her surprise.

"A horse," Graine repeated. "Teach me to ride."

"Why would you want to learn to ride a horse?"

"Must there be a reason?"

"Well no, but—"

"I know what I said. I told you, I take it all back."

"I can't imagine what has changed your mind," Grace said.

"Mr. Miller convinced me it would not be so improper for a lady to ride," Graine explained.

Now everyone was looking at Matt. "This is news to me," he said. "I'll not take the blame when her father hears."

Graine stared at them mischievously and said, "I have a pair of breeches that should fit."

"It's worse than I imagined," Will said. "One afternoon with my sister, and now you're wearing breeches?"

"Let it be our secret," Graine replied. "I don't want the whole town knowing…yet."

"I'd be glad to teach you," Grace said, "as long as your father agrees."

"'Twill be easier if he knows I have a good teacher," Graine said. "He thinks much of your family."

"We'll see after he learns of your desire to ride a horse," Will said. "He'll have us escorted off the grounds."

"He won't," she said. "He often asks why you never come to visit."

"Speaking of," Will said. "May I have a private audience with Graine?"

Matt smiled at him. "Certainly," he replied. "Grace?" He put his elbow out jokingly. To his surprise, she stepped over and put her arm through his.

When they were some distance away between the stable buildings, she stopped and said, "Graine's manner is greatly surprising. What exactly did you speak of?"

"It doesn't matter," Matt replied. He wouldn't have known where to start.

"Do you admire her?"

"Of course I admire her," he replied. "She's a wonderful woman."

"Romantically," Grace said.

"I'll be going back to Philadelphia soon," Matt replied. "Anyway, it's obvious that she's a perfect match for Will."

"Will has no admiration for Graine," Grace said.

"Yeah, she's only smart, beautiful, rich, and friendly," Matt replied. "He thinks she's totally repulsive."

"Those qualities are all a man desires?"

"They're a good start," Matt said. "Your brother has spoken highly of her, besides." He motioned around to all the buildings. "After seeing all this, there aren't many men who could support her...certainly not me."

"So you'll never be wealthy, Mr. Miller?"

"I'm a drunkard. You know that."

"I'm serious," Grace replied.

"I'm not sure why my being wealthy has anything to do with anything," Matt said. "Are riches all a lady looks for in a man?"

"No, but a lady does desire a man with the means to care for her family. I didn't make the rules by which we live."

"I plan on being wealthy," Matt said, "but I'm not now. There'll be a lot of uncertainty in my life over the next few years. The last thing I need is to take care of a lady."

"So you'd have no interest should the perfect lady come along?"

"Not at all."

She had a flirtatious smile on her face. "I don't believe you."

Matt ignored her. "I was hoping we could leave with enough time for my riding lesson."

"There should be time," she said. She let go of his elbow and headed toward the picnic, then turned back. "Are you coming, Mr. Miller?"

He had been standing there watching her walk.

# CHAPTER 30.

## MEN'S BREECHES

It was late afternoon by the time they returned from the Martins'. The farm was busy again now that men were arriving for the week. They were two horses short on the trip home, so Thomas had hooked one wagon behind the other. Thomas stopped in front of the house. Everyone climbed out, and then he continued to the side of the barn.

Matt went to the hay barn to change, closed the door behind him, and slowly removed his new clothes. He took extra care to fold it all properly and then changed back into hiking clothes. A strange metamorphosis was occurring in his mind; he was beginning to think of eighteenth-century clothes as normal and his twenty-first-century clothes as strange. He looked at the hiking boots, which he had worn for the first time only weeks before his trip. Even they seemed foreign after wearing his new black shoes all morning. He hoped that the hiking boots would fit the stirrups when he had his riding lesson. Once dressed, he wandered to the stables early to spend time petting and talking to Thunder.

Grace was already there and acted surprised to see him. She looked stunningly beautiful. "You've come already?" she said. "Don't dare laugh!"

"Why would I laugh?"

"Are you blind?"

"I don't know what you want me to see."

She looked down at her pants.

"Oh, the breeches," Matt said. He hadn't actually noticed.

"I wonder where you're from, Mr. Miller," Grace said. "There's not a gentleman in Virginia that wouldn't notice a lady in breeches."

"Where I come from, it's normal."

"In Philadelphia?"

"It doesn't matter. All you should know is that it's not surprising." He stepped back to look at her. "They're cute. You do look a little like a boy." She didn't look anything like a boy.

She hit him lightly on the shoulder.

"Will I get up on the horse today?" he asked.

"Yes," she said. "I want you to lead him into the ring. I'll ride Silver Star."

Matt took Thunder to the practice ring and then returned for the saddle. The dog wandered up and barked for Thunder's attention. The horse shook his head and made a loud *thpwaft* in the air. Scout scooted under the fence and trotted alongside the horse, making them look like they were going to do circus tricks. The more Matt watched them, though, the more moved he was by the beauty and grace of both animals.

Grace arrived proudly on her horse. Silver Star was a mottled silver stallion of medium size with rippling muscles. Silver didn't quite describe him, since he was more than one color, with patterns of dark streaks that painted his body and made him look like he was constantly in motion. His mane was silver-black and longer than most of the other horses', and meticulously trimmed. Grace looked like a druid priestess sitting atop her horse in front of the gate to the ring. Some of her hair had come loose from its bun and fell on her shoulders, framing her face. Matt couldn't help but laugh in wonder. She was a woman from a teenage boy's fantasy poster, a stunning beauty atop a gleaming horse. *The Horse Princess.*

"What are you laughing at?" Grace said.

"He doesn't hold a candle to Thunder."

She frowned. "The dog has to go. He'll distract the horses." She looked at Scout and pointed to the fence. "Out!" The dog glanced at Matt for confirmation.

"You gotta go, boy," Matt said. Scout stooped and slinked to the fence. He stopped and turned to make one last plea. "Hey, if it was up to me," Matt said. He put his hands out in a "what can I do" motion. "Lady says you gotta go, you gotta go."

"I don't know what's with you and that dog," Grace said. "We usually never see him."

"It's my charm," Matt replied.

"Not likely." She didn't let him answer before she continued. "You must learn to saddle him properly," she said, hopping down from Silver Star. They spent the remainder of the afternoon with Thunder, learning the basics of mounting, guiding, and riding him around the ring. Sometimes Matt would follow Grace's lead and other times she'd give instructions to him as she stood off to the side. Riding wasn't as hard as Matt had predicted, but after two hours in the saddle, his butt was sore and he was glad to be done.

They walked both horses to the pasture. Grace helped Matt remove Thunder's saddle and then they let him through the gate, watching as he trotted away. "I'm taking Silver Star for a ride," she announced. "I'll return before supper." She mounted her horse, twisted around, and galloped away towards the gate of the farm. Matt saw a few men glance up as she rode by, but they quickly went back to work. Matt grabbed the saddle, took it to the barn and eased it onto an empty wood frame.

Thomas was across the barn, working with some leather straps attached to a metal bridle. Matt walked over. "You'll need to buy a saddle for your horse," Thomas said.

"A saddle doesn't come with every horse?" Matt asked, only partly joking. He hadn't thought about the additional expense.

"If you pay enough, we'll include anything you desire," Thomas said, smiling.

"I was hoping it was buy one, get one free," Matt explained. He wanted to intrigue the man enough to negotiate a reasonable price for both horse and saddle.

"Interesting proposal," Thomas said. "Most people would charge twice as much for the first item, I imagine."

"What's a saddle cost?"

"About three pounds for a new saddle in town," Thomas said. "We'll sell you one at cost once we see your money for Thunder."

"You wouldn't sell him before I have the money."

"Someone could come any time," he said. "I don't believe you can afford the seventy pounds anyway."

"Seventy pounds?" Matt said, incredulous. He tried to get a read on the man to see if he was joking, but Thomas's expression was blank.

"Too much?" Thomas asked. "How much do you think he's worth?"

"I have no idea,"

"You've already sold four horses," Thomas said. "How much?"

"I'd ask forty-five pounds," Matt replied, "but would sell him to a guy like me for around thirty-five."

"Do you have thirty-five?"

"You know how much I have," Matt said. "You're keeping my money in your strongbox."

"I'll hold him for a five-pound deposit," Thomas said. "I'll expect another thirty pounds when your business is complete in Richmond."

"It's a deal," said Matt.

"How goes the instruction?" Thomas asked.

"I've learned a lot," Matt said. "Grace loves to ride."

"Both Grace and her sister had horses in their blood—" Thomas said, and caught himself. "Have? Had? It makes me proud to see my children with the animals."

"What about Will?" Matt said, chuckling.

"Will favors the enterprise. The animals are a means to an end," Thomas said. "Grace would be happy to keep every animal."

"She rode her horse today," Matt said.

"We try not to discuss Silver Star around Will," Thomas replied. "My son can't resist mentioning buyers. What're your plans after harvest?"

"Once my ring sells," Matt replied, "I'm off to Philadelphia to restart my life."

"Buying a horse is a worthy start, then," Thomas said. "You should arrange for time during harvest to continue instruction. Arrange it with Uncle."

After leaving the barn, Matt walked over to the corral to inspect Joshua, who was eating happily. "Five more days to go, boy," Matt said aloud. The infected skin was no longer red, and the sores were starting to lose their scabs and heal over. Matt looked at his handiwork one more time, glanced up at the sky and whispered thanks, then headed to the common. His plans were to sit with the workers and have an ale. He felt like he should be tired after everything he had experienced on his second Sunday in 1762, but he was nothing less than exhilarated. It had been a good day.

# CHAPTER 31.

# I DON'T KNOW

Matt never wavered from his conviction that picking corn was misery. His first week of farming had been almost all about learning new skills. In this second week, he learned all there was to know about picking corn in a single morning, and after that, it turned into a monotonous bore. He had done his best to look at the stalks in wonder on that first day, using Thomas's metaphor that they represented the affairs of men. By the second day, though, the stalks looked the same, like people walking by in a shopping mall. By the third and fourth days, the only metaphor he could accept was that he was in jail, and the cornstalks were fellow inmates.

What Matt did look forward to every day was riding. He'd join Grace about two hours before dinner, and they would talk about horses as they brushed and saddled their animals, then spend the remainder of their time in the practice ring. It was already Thursday, and Matt was about to begin his fifth lesson. Grace had told him to plan on a long ride in the countryside.

After four lessons, he had learned a lot about horses, but he had learned even more about Grace Taylor. Matt liked how she'd start talking about horses, then let the topic morph into something else as they rode together. Horses were an integral part of colonial life in many ways, so Grace was able to build a detailed picture of her family, her community, and her culture by relating them to her experiences with the animals and the people that bought them.

The riding lessons required Matt's complete concentration and he often entirely forgot about Grace the woman; he mostly saw her as a friend and instructor. He tried to ask her provocative questions about her experience with horses, and he was usually intrigued at her answers. Someone had once told him that the best way to learn any skill was to ask a person to narrate it from their perspective. It was proving to be an excellent way to unravel the mystery of this beautiful farmer's daughter.

"What do you think about when you first get up on a horse?" he had asked her.

"That I'm glad to be with my fellow," she said. "And that I don't want to betray his trust."

"What trust?"

"He'll do whatever I ask," she replied. "I must not ask him to do something that may cause harm." She paused. "I should not squander his time, either, so I must remember that."

"How can you squander a horse's time?"

"He must be a horse and do what horses do."

"You talk about horses like they're people."

"They're God's animals, and He put them on this earth for a purpose, like men."

"So you're doing God's work?" Matt asked, smiling.

"It's not as grand as that," she said. "Horses deserve respect, much like that dog of yours." She nodded to Scout, who had started the habit of watching them at lessons every day. "They feel the admiration and respond in kind."

"Are you ever upset when they're sold?"

"There are times," she said. "A few I haven't cared for."

This surprised Matt. "Which horses didn't you like?"

"The twins," she replied. "They were too distracted with each other to have time for a human rider. Mr. Hancock always drives a coach, so they were a perfect choice."

"Do you like Shadow?" Matt had asked. "I heard only your father is allowed in with him."

"This much is true," she said, smiling, "but it's for Shadow's safety, not mine."

"Shadow's safety?" Matt asked, puzzled.

"I struck Shadow square in the face this year," she said. "Father won't allow me near him because I wouldn't pledge not to do it again."

"You hit a horse in the face?"

"As hard as I could," she said with a grin.

"Why'd you do that?" This was the most entertaining thing he had learned about Grace thus far.

"Anyone who spends time with Shadow wishes to strike him," Grace said. "He deserved it."

"Did you hurt him?"

"Not as much as I desired," she answered. "I split his lip. He remembers when he sees me."

"Isn't that a little extreme to strike a horse in the face?"

"I don't know," she replied. "Is it extreme to strike a man in the face?"

"Not if you're defending yourself," Matt admitted. "It's entirely justified."

"We're usually defending ourselves around that pernicious animal," she said. "So you have your answer."

**\*\*\*\*\*\*\*\*\*\***

Grace arrived for their Thursday lesson dressed in a pair of blue breeches, a white shirt, and a dark Spanish-looking hat with a chinstrap, like something from a Zorro movie. It was hard not to be impressed as she trotted up on her silver stallion.

"You ready to ride?" she asked. Matt detected a mild taunting in her voice.

"Can you keep up on your tiny horse?" he replied.

"I greatly desire to learn," she said confidently.

Matt had been standing beside Thunder, talking to him. He touched his head and whispered in his ear, "Let's show the princess what we can do." He mounted Thunder and pulled back on the reins to make him rear up slightly. He smiled proudly at Grace and looked down at Scout. "You coming?" The dog's ears perked.

"He should stay," Grace said.

"Afraid he'll spook your horse?" Matt teased.

She laughed and without a word pivoted Silver Star towards the gate and moved quickly out onto the road. Matt nodded to Scout as he clicked his heels against Thunder, and the three of them joined the chase. It was the first time Matt had ridden Thunder outside the ring. Grace cantered some distance before stopping to let them catch up.

"I know you believe you've mastered riding, Mr. Miller," she said, "but I'd proceed slowly."

"I don't feel like I've mastered anything," Matt replied. "Don't forget that I'm still paying you to teach me."

"You'll get what you paid for," she said, smiling. "Today, you'll learn to gallop."

"Do I need to gallop?"

"He'll be a weary animal should you never let him run. Do you wish him to grow melancholy?"

"I was joking. Galloping seems way beyond my skill level."

"You've done a canter in the ring. Let's go for a faster canter until you both are comfortable, and then we'll attempt the gallop."

Grace and Matt rode side by side, with Scout following close behind, looking up in anticipation. Matt rose up in the saddle as Grace had shown him, and then tried to move Thunder out of the trot as quickly as possible. Easing into a canter smoothed the ride out enough that Matt didn't feel like the horse was trying to bounce him off his back. They moved at a fast pace for a while and then Grace slowed.

"Now," she said, "begin to canter, then rise up from the saddle and squeeze your legs as you move your body forward. Grab onto his mane and guide him into a gallop. Can you do this?"

"I think," he said. "Yes." It all seemed too out of control. He focused hard on looking confident.

"If you start to lose him, ease back," she said. "I don't want him injured."

"What about me?" Matt said seriously.

"Your broken leg will heal. Thunder would have to be put down." She moved Silver Star forward and waved for Matt to follow. Matt kicked Thunder into motion and caught up to her so they could ride in parallel. "Now," she yelled. "Ease him into a gallop." Matt squeezed his legs together and leaned forward. The horse only cantered faster. Grace trailed behind them. "No," she said, "a gallop."

"I'm trying," Matt said. "He won't go any faster."

"Squeeze and move forward."

"He won't go," Matt repeated. No matter what he did, Thunder refused to run.

"Follow me," she directed as she jumped Silver Star ahead of him. Grace increased her distance, but it was evident that Thunder didn't want to gallop. She was gradually moving away. The dog noticed and shot forward, running ahead after Grace. Thunder saw this and went off like a rocket. As the horse broke into a run, pounding Matt on his back, it took every ounce of Matt's twenty-first-century composure not to throw himself off and be saved. Horse and rider moved in a state of controlled chaos as they chased Grace and the dog. They ran for nearly a minute before Grace eased Silver Star to a walk in front of them and Thunder throttled back to match their pace.

"You did it," Grace called. "Try again."

"If you want," Matt replied. She took him at his word and kicked Silver Star forward. This time, Thunder didn't need the dog to respond to Matt's command. They were galloping, chasing Grace on the country road. She continued to lead, moving away for another minute before she slowed, allowing Matt to catch up. Both horses breathed heavily as they walked together.

"Thunder will be pleased that you let him run," she said. She looked at the sun in the sky. "We should probably return."

"Do we gallop back?"

"No," she said. "You never want him running to his stall or he'll get the notion that he can run back whenever he pleases. Take your time so he knows who's in control."

They turned their horses and started home. The dog saw this and took a position trotting in front of them, looking back regularly to make sure they were keeping up. They rode in silence down the dirt road for a while, enjoying the solitude and the warm sounds of the countryside, until Grace finally spoke.

"Let me show you something," she said, pointing to a trail ahead of them that intersected the road. She broke into a run out in front of him and was around the corner before he could think of following. The dog looked up at him, asking permission.

"Go ahead!" Matt said. He kicked Thunder lightly as they rounded a bend and the horse cantered forward. The dog moved to the side as they passed him. Grace was now a fair distance away. The dog saw this and sprinted ahead. Much to Matt's dismay, Thunder galloped after him. Matt grabbed the horse's mane and put his head down, and they soon caught Grace.

"I hoped you'd arrive eventually," she said.

"Strange coming from a woman who told me to be careful," he said. "How fast do you want me to ride?"

"Faster than a lady on sidesaddle." She kicked and Silver Star leaped forward. Matt followed her out of the woods to a clearing that overlooked a river. "The James River," she said, pointing.

"It's amazing," Matt said. They were in a place where the river narrowed and fell over a raised cliff as a waterfall and then into a deep pool, which emptied as it continued downstream.

"I often came here with Kathryn when we were girls."

"Everybody talks about how wonderful your sister was," Matt said.

"Beautiful...talented, friendly...graceful," Grace explained. "I felt ugly and awkward around Kathryn."

Matt stayed quiet, mostly because he had no idea how to respond. He couldn't imagine Grace feeling anything but beautiful.

"When I speak thus," she said, "you're obligated to reply, 'Grace, you possess those qualities as well.'"

Matt laughed. "Your name's Grace," he said. "I guess you have that covered."

She hit him on the shoulder lightly. "Not what I was seeking, certainly." She turned on Silver Star and was off toward home, with the dog following close on her heels. They raced until they could see the gate and she slowed Silver Star in front of him. He trotted up beside her. The horses breathed heavily, sounding like locomotives.

"You learned to gallop."

"What's next?" he said back, smiling.

"We should take a few more rides in the country to shake out the cobwebs. It's been a long time since he's run." She motioned to Thunder. "He's happy."

Matt reached down and patted Thunder's neck.

"See you in a moment," Grace said. "It's the last supper before the men go back to town."

Matt watched her trot away in her Zorro hat and smiled in wonder at a God that could have made such a gorgeous creature. He patted Thunder's neck again and said, "Let's get you put away." He hopped off, removed the saddle, and walked the horse until he had cooled. He then let him in with the other horses and went with Scout to wash for dinner.

"We should go see how Joshua is doing," Matt said to the dog as they left the hay barn. The dog's ears went up and he trotted out ahead to lead the way to Joshua's corral. When they rounded the corner, they saw that Grace was already there checking his back, and so Matt stopped outside the gate of the corral. "How's he look?" Matt said.

"There's even hair starting to fill in the bare spots," she replied. "Do you want to see?"

"I believe you," he said. "You coming to supper?"

"Wait," she replied, motioning with her forefinger. She reached up and rubbed Joshua's cheeks with her hands, then pulled his head down and kissed him on the nose. She whispered something to him, then walked to the gate where Matt was standing and let herself out.

"Father wants to know when we should let him back in with the others," she said.

"I thought he's the one who would decide."

"He said 'tis the apothecary's decision."

"I want to wait a full ten days," Matt said. "Does it matter?"

"He's growing fat," she replied. "He should be out running."

"One more day in the corral and then you can let him in with the others," Matt said. "Don't wash the medicine off until Monday to be safe."

Grace was walking beside him now between the barns. "Fine."

"Simple as that?" Matt replied. "Fine?"

She stopped abruptly to face him and Matt was forced to stop with her. "What about *fine* vexes you?" she asked. There was a good-humored grin on her face.

"Usually you've some smart reply," he replied. "I'm used to more is all."

She looked up at him silently and stepped forward, and the next thing he knew, she was kissing him. When it was finished, he leaned away from her in surprise. It all had happened so fast he had trouble remembering the kiss.

"What was that for?" he said.

"I don't know," she replied.

"Didn't you just kiss the horse?"

She stepped close again, slid one hand around his waist, put another on his neck, and pulled his head towards her. He had no trouble remembering what her kiss felt like the second time; it was deliberate, firm, and wet, and he had barely kept his knees from collapsing beneath him. Afterwards, she stepped slowly away and walked toward the house as he watched her. She turned back and smiled. "Are you coming?" she said as she rounded the corner out of sight.

Matt looked down at the dog. "That really happened...right?"

# CHAPTER 32.

## HARVEST'S END

The men were festive and happy now that the end of harvest was near. Matt was standing in line with Charles and Zachariah, waiting for what looked to be a substantial feast. The two Richmond men talked excitedly about having been offered steady work, and they speculated as to their future prospects. Hiring full-time employees was new for the Taylors, and they would be working out the details as they went.

Despite his desire to share their excitement, Matt couldn't help but be distracted as he regularly stared towards the farmhouse, trying to catch a glimpse of Grace.

"You once mentioned you would return to Philadelphia," Zachariah said.

"As soon as my business is complete in Richmond," Matt replied. "I'll learn more in town tomorrow."

"What kind of business?" Zachariah asked.

"I have something for sale at auction up North. I'm waiting for that."

"You said you'd buy a horse," Charles said. "Have you found an animal?"

"I've put a deposit down on Thunder," Matt said. "The large chestnut here on the farm."

"You bought a Taylor horse?" Zachariah asked, surprised.

"I put a deposit down. The auction should bring enough money to pay for the horse."

"I've seen Thunder. He's a beautiful animal and could comfortably carry a man my size," Charles said. "Next thing you know, you'll be courting the Taylor daughter." He smacked Matt on the shoulder in good humor and laughed. "Don't forget me when you're a wealthy merchant."

"And don't forget me, either, when you're one," Matt said.

This made Charles smile. "It's a deal. When do you go back to Philadelphia?"

"Another week before I get my money," Matt said. "I can't sleep in the barn for much longer without Thomas Taylor chasing me out with a musket."

"Good evening, gentlemen," said Will, walking up behind them. Matt glanced past Will to see that all the Taylors had arrived. Grace was standing next to her father, talking. She met Matt's gaze but then quickly turned back to her father.

"Mr. Miller!" Will had a sly smile on his face when Matt finally looked at him. Matt had no idea how many times Will had actually said his name. "David mentioned that you should speak with him."

Matt nodded and smiled. "I'll go."

"Wear your blinders," Will said, laughing as he put his hands up to his eyes. Charles and Zachariah looked at Will, puzzled.

Matt rolled his eyes. "No comment." He turned from them and walked over to where David was tending the fire.

"It's the final evening for most," David said when he saw him. "You needn't work in the stables tonight. This may be the last you see many of these men." Matt had worked the last three nights in the stables cleaning out stalls to make up the time spent riding. "How are the lessons coming?"

"I learned to gallop today," Matt replied.

"Grace is glad to ride again," David said. "You get full credit."

"I'd never have gotten involved if it weren't for our conversation," Matt said.

"Thomas said you'll be letting Joshua back in with the other horses."

"One more day in the corral," Matt said.

"How much would you charge for something like that in Philadelphia?"

"Four or five pounds," Matt said, "for medicine and time spent."

"Five pounds, then?" David asked.

"Zero pounds for you. I'd have paid my own money to see Joshua cured. Grace, Jonathan, Jeb, and I worked together."

"You healed that horse for Grace," David said simply.

Matt thought for a second and realized he was probably right. "Maybe I did," he replied. "Does it matter?"

"If what I observed is true," David said.

"What've you observed?" Matt asked.

"That girl's falling in love with you."

Matt's stomach jumped with both surprise and satisfaction. He thought for a moment. "Would that be bad?" he asked. "I think I may have loved her since the first day we met."

"You seem to be doing everything possible to seduce her," he replied. "She's riding, the horse is cured, and she's speaking with Graine Martin." He threw his hands up in front of him in a "What's next" gesture.

"Maybe I have been doing it on purpose," Matt said. "Don't men spend their entire lives trying to make women happy?"

"You've got a secret you're not telling us," David said, "and I don't like it."

"What would it take?" Matt asked.

"I've thought about this much since our ride into Richmond. Only time builds trust. I've known her other suitors for years."

"It's really not an issue," Matt said, irritated. "I'll be leaving soon. I know enough about Richmond society that I'd never be accepted."

"Hold on, boy!" David said. "I didn't say I wanted you off the farm!"

"Do you think I'm so bold as to think that Grace Taylor would marry a simple farmhand?" Matt exclaimed.

David looked back and laughed. "Son, I'd not put it past you." His laughter was genuine, and it broke the tension that had built up between them. "You'd better tell her." He paused for a moment and said, "We all know you're not a simple farmhand."

"What do I say?" Matt asked.

"Tell her your intentions," David replied. "What's there to lose?"

"Easy for you."

"Now's your chance," David said, glancing over Matt's shoulder.

Matt turned his head to see Grace walking into the common. "I have to do this now?"

"Before you lose your opportunity…or your fortitude." David laughed again and gave Matt a warm smile, clearly looking forward to having some fun at the younger man's expense.

# CHAPTER 33.

# COURT WELL, MR. MILLER

An overwhelming sense of déjà vu debilitated him as he made his first step towards Grace. Strings of white light from his dreams pulled him to a place he had already been or seen. Flashes of the future jittered in his head and he had to stop walking to force himself out of what seemed like a waking dream. Fortunately, the visions and strings of light faded by the time he reached Grace.

"Good evening, Mr. Miller," she said. "Come to help?"

"No," Matt said with a laugh he knew sounded tentative.

"I must take these pots to the well."

Matt shook his head, saying, "Well, I thought we could—"

"Hold these," she interrupted as she began loading pots into his arms.

"Okay, but—"

"These too," she said, adding more. She grabbed the others and said, "Follow me." He hurried behind her, the pots clinking as he tried to match her pace. He could barely hear above the clatter.

"Do you plan to ask Father if you can court me?" Grace said over her shoulder.

"Do you want me to ask your father if I can court you?"

Grace stopped and turned, short of the well. Matt's momentum almost carried him into her. He stopped in his tracks, holding his moving and clanging pots. "I kissed you!" she said.

"I wasn't sure what that meant," Matt replied.

"What else could it mean?"

"It was only a kiss, and I—"

"Only a kiss! Do you kiss many ladies?"

"I wanted to talk to you and ask—"

Grace threw a pot at him that clanged off the ones he was holding, then turned around and marched to the well.

"Grace, come back," Matt called. He reached down, hooked the pot she had thrown, and hurried to follow her.

"I have naught to say to you, Mr. Miller. A kiss means the same from everyone. It's the first time I kissed a man since I was fourteen and I can't believe I wasted it on you."

When they reached the well, Matt waited in fascination for her to finish what was turning out to be an epic rant. Eventually, he relaxed his arms and all the pots dropped onto the ground. He closed his eyes against the clatter. "I'm in love with you," he finally said. Then, remembering the rest of her comment, he asked, "What man did you kiss when you were fourteen?"

Grace picked up a pot and handed it to him. "Scrub."

"If I ask your father and he says yes, then what?"

"I admit, I've not thought that far."

"I'll have to go back to Philadelphia."

"Why?"

"To start a business."

"You can't practice your trade in Richmond?"

"It's not big enough for two apothecaries."

"There's work here."

"You'd be happy marrying a farmhand?"

"How long will this business of yours take?"

"How long will you wait?"

"It depends on who else comes to call," she said, "and how well you use me until you leave." She wrinkled her nose. "Court well, Mr. Miller."

"I'll talk to your father," Matt said.

"Will you travel into Richmond on Saturday?"

"Yes," he replied. "Why?"

"Father is going. You can ask then."

"Maybe. This is my deal, though."

Grace reached for another pot and they continued to wash in silence, frequently glancing at one another. Their eyes would meet for a moment and then, as if there was some signal, they would go back to their task. Matt could already feel that his courtship was predestined. The glances and nods, even the pots passed between them, represented their agreement. She picked up the pots, handing them to him, and then he followed her down into the kitchen, where he helped her hang them on the rack one by one. "That's it," she said when she was finished. "I'll come to supper in a moment." Matt lingered there, drinking in her beauty. She made the motion to step toward him and he met her halfway. She melted into his arms and they kissed long and hard. When she finally stepped away, he was dizzy.

*********

He shook the haze from his head as he walked from the house, wondering how she had bewitched him but glad that she had. He meandered to the common on ground that felt soft beneath his feet, like he was walking on cotton. David was the first to see him. "I saw her throw a pot at you," he said.

"That was only the start of the negotiation," Matt replied. "I promised to ask her father for permission to court her."

"What's your plan?"

"I don't have one yet," Matt replied with some dismay.

"Sometimes you make me believe you're not a scheming villain after all."

"In the unlikely event that I'm not," Matt said, "what should I say to her father?"

"Ask him if you can court his daughter," David said. "What else is there?"

"What does it mean, to court someone?"

"You've never courted a lady? A handsome fellow like you?"

"Not a good Christian lady," Matt replied. "For what specifically am I asking permission?"

"To spend time," David said. "Kiss her hand, that sort of thing."

"I'm already spending two hours a day with her on horseback," Matt said. "She'd laugh at me if I tried to kiss her hand."

"Formal visits are out. You're already here," David answered after thinking. "I'd ask for permission to write to her when you're gone. You want permission to escort her to the Martins' party on Sunday."

"Party? What party?"

"A picnic. Charles is coming out to mind the farm. Even I'm going."

"Was I invited?"

"Graine mentioned you specifically on the invitation," David said. "I can only guess how you managed that. It's most people in the church, including the Reverend Michael. They've his blessing for dancing and games, even though it's the Sabbath. He likes parties more than anyone."

"Dancing?" Matt said, surprised.

"Ask Will to teach you before he leaves," David said. "He's the best dancer in the family, after the ladies."

Matt frowned. "Dancing lessons?"

"You don't want her out there dancing with someone else, do you?"

Matt saw Grace arrive at the common. She was glowing. His lapse in attention wasn't lost on David. "You think you're worthy of my niece?"

"I could be," Matt replied, uncertain.

"Good evening, gentlemen," Will said, walking up to them. "What do you two speak of so intensely?"

"I'm looking for someone to teach me to dance," Matt said.

"I'm a rather good dancer," Will said. "I don't know if David told—"

"Already did," David exclaimed, "so there's no need to boast."

"I'm merely saying," Will said, "that I've mastered the fashionable dances."

"I don't know what Graine sees in you," David said.

Will backed away, putting his hands up. "Come on, Levi!"

"I'm serious," Matt said. "I need a dance teacher."

"After dinner tomorrow," Will said. He and David began mock fighting, and laughing uncontrollably. Will glanced at Matt with his hands up as he deflected David's jabs. "The Martins' party?"

"Yes," Matt said, resigned. He felt he was living some predetermined story that everyone had read but him.

"You'll be ready," Will said as he backed away from his uncle.

Thomas walked up during their horseplay. "What're you doing?"

"Fighting Levi," Will said, "like Mr. Miller." He picked his foot up, pretending to kick his uncle.

"Act like the owner's son," Thomas said. He only seemed to be partly joking. He looked at Matt. "Mr. Miller, Grace mentioned you're interested in riding your horse into Richmond to get some experience around town."

Matt had no idea what he was talking about, but decided to go with it. "Yes, sir. I'd also like to visit Jacob Berkley." He saw a wide grin fill David's face.

"I've business in town," Thomas said. "I'm sure we can match our schedules."

"Yes, sir," Matt said. "I look forward to the trip."

"We can talk at supper tomorrow," Thomas said. He looked at all of them. "Don't stay out too late."

"You sure you don't want to help me defeat this wicked villain?" Will said, making a stab at David, who stepped back.

"Everything's a joke," his father replied.

"I give up," Will said. "You can be Mr. Miller and I'll be Levi in an epic battle."

"Good evening, gentlemen," Thomas said, frowning. He turned and walked toward the house.

"What's with all that 'yes, sir' talk?" Will exclaimed. "If I didn't know better, I'd say you endeavor to court my sister." Matt had no idea what to say. Will scrutinized them. "'Twas a jest," he said, confused by their serious faces.

"Can I court your sister?" Matt said.

"You're serious," Will replied. "What does she say?"

"She said I should ask."

Will glared at his uncle. "Did you know of this?"

"I saw her throw a pot at him," David said. "That's the extent." Will gave him a puzzled look.

"Can I court your sister?" Matt asked again.

"If you use her well," Will replied.

"How else do you think I'd *use* her?" Matt said. He had trouble even saying the word *use*, considering its negative connotations in his own time.

"You make her smile. It's a welcome change," Will replied. "You have my permission, for all it's worth. Father's the one you must convince."

Matt looked at David. "Her uncle, too."

"You'll have your chance," David replied.

\*\*\*\*\*\*\*\*\*\*

The Taylor family began saying their goodnights. Grace approached the three of them with a smile and only briefly glanced at Matt. "I'm off to bed," she declared. No one said a word and there was an uncomfortable silence. "Good evening," she said loudly, trying again for some response.

"Anything you should tell me?" Will asked.

"Naught, dear brother, until Mr. Miller has had his transaction with you and Father." Matt could see the surprise on Will's face at his sister's very casual reply.

"He spoke of this to me already," Will said.

"And?"

"I said yes."

"One more, then," she said, smiling. "Goodnight, gentlemen." She turned and was gone.

"Zounds!" Will exclaimed. "Not a frown. I've never seen her thus."

**\*\*\*\*\*\*\*\*\***

Matt walked by himself to the barn. He was now resolute in the feeling that he could no longer ignore the men who were trying to rescue him. He wanted the issue settled, so when he entered the barn, he turned on the phone for the first time in days and sat there waiting. The text came exactly on time.

"Are you there?" it read.

Matt typed, "Yes."

"Could not contact you at the agreed time. Some problem?"

"Here now."

"The portal will open at your original entry point in 48 hours. Can you get there?"

"I'm not coming. Rescue one of the others."

"You can't stay in 1762. We don't want you changing the future."

"Side effects of the portal are too dangerous. I'm staying. The timeline will be fine."

"It's an order. You must be there."

"I'm not military. Tell my dad, Andrew J. Miller, I'm okay and I love him."

"Contact you in 24 hours to discuss."

Matt shut the phone off, dimmed the lantern, and crawled into bed. He fell asleep quickly to the sound of the dog's snoring.

# CHAPTER 34.

## WHAT'S UP?

It was all about corn again on Friday. The men who weren't working out in the field were back shucking. Matt spent the morning picking ears and throwing them into one after another wagon. Some of the wagons would be taking corn into town for sale, others to drying towers to be dehydrated, and yet more would be mixed with grain and hay to make silage. The crop had so many uses that it should have been fascinating, but all Matt could do was watch the sun until it was time for his riding lesson.

His thoughts wandered through the possibilities for his future. There were moments when he wished the Taylors had never found him and that he had never met Grace. He considered how easy it would've been to wander into town, sell his ring for thirty pounds, and get passage back to Philadelphia. Even now, holding the tiger by the tail, he was tempted to let it all go and take the path of least resistance. As he walked along, picking corn, he tried to remember his visions and dreams. Was there a detailed vision of Grace there somewhere? He was seeing visions of the future every night now and sometimes even during the day, but none seemed clear enough to guide him.

When Grace rode up for their daily lesson, she was wearing a completely different pair of pants. "You own a lot of breeches," he proclaimed.

"I knew they would nettle you eventually," she replied.

"You own more pairs of pants than I do."

"Buy more."

"I can only carry so much on horseback. I need to travel light."

"Thus I've heard. You ready to ride?"

"Thunder can't wait," Matt said. The truth was that the saddle yesterday had made him sore. He was hoping today's lesson would consist of a leisurely ride in the countryside.

"Most of the men are on their way back into town," she said. "I want to give you practice riding between barns and with people still about. Mount up." Matt put his foot in the stirrup and lifted himself onto the horse. Grace purposely led him into tight situations so he could practice backing out. A couple of times, farmhands coming out of doors or around corners surprised them; Thunder did startle when something unexpected appeared in close quarters.

"Sometimes dogs spook horses," Grace said, looking at Scout, who was following close behind. "Probably not an issue in this case, but best to be aware."

Matt tried to keep Thunder away from buildings, but the horse didn't seem to care if he sideswiped walls or fences. "Keep him away from obstacles unless you want your legs smashed," Grace said.

"I'm trying," Matt replied.

"You're letting him go wherever he chooses," she scolded. "Control."

"I'm trying," Matt replied, frustrated. Thunder had dragged him across a fence and his leg had been jammed in between horse and wood.

"I imagined as much," she declared. "You've been letting him do the work."

"What'd you mean?" Matt said, focusing on steering the horse away from another barn wall.

"Strike him on his side if he goes too close," she said. "Show him!"

They did it again, and this time Matt smacked Thunder every time he neared a fence. After the first couple of near misses, the horse learned to stay away. "You're right," Matt said.

"You need not be gentle," Grace said. "Not too long ago, Thunder's ancestors were fighting wars." Matt was following her as she threaded her way through the outbuildings. "He's smarter than most horses," Grace said. "Smart horses get lazy. He'll test your resolve."

"I get it," Matt said. "It's like everything else; he has to respect me."

"You respect each other," Grace replied. "Let's ride the country roads."

As she said this, she made Silver Star rear up on his hind legs and turn completely around to head back to the farm gate. Matt shook his head, thinking how reckless she seemed on horseback. He turned Thunder and trotted after her. When she reached the gate, she broke into a gallop. He groaned when he saw this, thinking of his sore butt.

When Thunder reached the open road, Matt tried to make him gallop, but the horse refused, again. He could see Grace riding like she was in the Kentucky Derby. Matt reached back and smacked Thunder's rear as hard as he could, and Thunder jumped forward. "Ha!" Matt said. "Finally, you scoundrel." Even with his long strides, it took a lot of pounding for him to catch Grace, who had been galloping Silver Star full-out. She slowed when he reached her.

"Arrived finally?" she taunted.

"What the hell was that about?" Matt exclaimed. The lower part of his body felt like someone had beaten him with a bat.

"Such language!" she replied. Matt could barely make out her words above the loud breathing of the horses.

"You're trying to kill yourself," Matt said.

"You know naught about horses," Grace said. "There's nothing wrong with running them hard." They walked side by side now and had traveled a fair distance.

"That's a bunch of crap," Matt replied. "You're trying to kill yourself."

"Should I die, I'll be doing what I choose."

"You've your whole life to ride your horse."

"Until a man says I cannot," she replied. "I'll be a wife soon, never to ride again."

"You think this even about me?"

"What else would you do?"

"Women are responsible for themselves where I'm from."

"They must do what their husbands command!"

"I wouldn't know the first thing about commanding my wife."

"You think we'll get married?"

"What do you mean?"

"Father will never agree," Grace said. "My betrothal will be to a wealthy Virginian."

"We have to try," Matt said.

"He's worked all his life so his children could marry into the South's best families," Grace said. "I'll never have the luxury to marry a man for love."

"Do you love me?" Matt asked.

"Maybe," Grace said.

"Maybe?"

"What do you expect?"

"I expect you to wait," Matt replied. He surprised himself with his calm focus. "I'm leaving for Philadelphia soon. I'll be back with an offer that your father can't refuse."

"Waiting's not a problem," she said. "I'm still unmarried, as you may have noticed."

"One day, you'll have your own stables and you'll be teaching our daughters to ride."

"Do you pledge this?"

"I pledge this," Matt replied.

"And you'll not gripe when they wear breeches to ride their horses?"

"As long as they don't look like boys," Matt said, joking.

"Our daughters will resemble me," she said. "I don't look at all like a boy." She kicked Silver Star and trotted off ahead. Matt followed her back, staying close enough to watch her on the horse as her body swayed from side to side. He wanted her then, like nothing he had ever known.

They went their separate ways as they entered the gate. Matt knew Grace had as much to think about as he. He guided Thunder to the corral, dismounted, and pulled the saddle off to take it into the barn. Jonathan appeared from nowhere, saying, "Good afternoon, Mr. Miller."

"What's up?" Matt said. He had grown used to the boy appearing out of nowhere.

The boy looked up at the sky. "The clouds, I should think," he replied.

"That's funny," Matt said. "'What's up?' is like asking someone what they're doing."

"Father sent me to tell you that dinner would be in half an hour if you must *wash* up," Jonathan replied.

"I'll be over in half an hour," Matt said. Matt expected the boy to leave, but he sat down instead. "Now what's up?"

"You glad harvest is over?" Jonathan asked.

"Mostly," Matt said. "It was hard work."

"You leaving soon?"

"I can't live in this barn my whole life."

"Too itchy with all the hay?" Jonathan gave him a giant smile.

"You know why."

"Your trade is in Philadelphia."

"No matter how much I want to stay," Matt replied.

"How will you marry Grace if you're in Philadelphia?"

"What makes you think I'll marry Grace?"

"She likes you. Everyone likes you." The boy hopped to his feet. "I must return to set the table."

"See you later, sport."

"It's Jonathan," he said as he was walking away.

# CHAPTER 35.

# TOOTHPASTE, PART IV

It had been nine months since the accident. The four physicists had been operating in relative secrecy since then, having received orders from Colonel Gabriel to keep working on a project that the United States government no longer sanctioned. During the day, they developed weapons-grade laser technology, but at night when the labs grew silent they delved into the unknown interactions between space and time.

They'd had probing calls on a few occasions from the Department of Defense, asking them to explain the strange energy signature detected coming from the laboratory every few months. Its magnitude was nothing like that detected during the original accident, so they explained the surges as an artifact of testing laser weapons. The lasers emitted no such signature, of course. The spy satellites were detecting the controlled particle stream required to open up a quarter-size hole in space-time in order to send a text message to Dr. Matthew Miller, the twenty-six-year-old American scientist who had been accidentally pushed through a wormhole to the year 1762.

Despite their many attempts, the physicists were unable to contact the other three people trapped in the field. They had an almost endless number of theories as to what had happened to them . . . they might be in a different time; their phones might be dead or damaged; they simply might not be interested in being rescued. The other possibility, which none of them wanted to talk about, was that being squeezed through a wormhole as it collapsed behind you could be fatal.

Nine months of experiments had given them a working knowledge of the relationship between wormholes and time travel, but because they had to operate undetected, much of their understanding had been extrapolated from the initial accident and by studying the miniature holes in space-time that they had been able to form in the laboratory during their communications with Dr. Miller. Despite their wealth of knowledge, most were uncomfortable with trying to pull Matthew Miller back to his original time. They debated it endlessly.

Jacob Cromwell's support waned as the project continued. He didn't see any value in contributing to the damage they'd already caused. He was the only one of the four physicists who had a child. As he watched his son grow from an infant during that year, he became increasingly reluctant to jeopardize his career. He no longer wanted to work on a secret project that seemed to have very little upside. "What if we leave him in 1762 like he wants?" he asked as they all stood around the reactor.

"We don't know how he might change the future," Brian Palmer replied. "We can't leave him there." Palmer had become the most enthusiastic of the physicists on the program. From his perspective, there was *only* upside to their covert research. He saw himself in the history books as the man who probed the relationship between space and time; the man who proved that time travel was possible. The only way they could disclose their discovery was to bring all four people back safely. Any other option would lead to an inquiry into their fate, and eventually the question of whether the accident caused their deaths.

"It might be true what he said about the timeline requiring him to be exactly where he is," Cromwell reasoned. "Either way, we can't force him to step into the hole."

"We have to fix the problem we caused," Palmer said. "Once we have a signal from his phone, we can open the wormhole right on top of him."

# CHAPTER 36.

## HAY BARN

Matt took his time washing for dinner. He put on clean clothes and walked to the house. When he entered, only Mary and Grace were in the kitchen. "Where is everyone?" he called.

"Letting Joshua back in with the other horses," Mary said. "You can join them if you desire."

"If it's all the same, I'll sit and rest."

He poured a cup of cold water and drank it like it was fine wine while he glanced down at Grace, who was helping her mother cook. It was one of those perfect moments, ripe with possibility. The feeling reminded him of graduate school, when he had worked in the laboratory. He'd spend a week planning a chemical reaction, gathering all the chemicals and supplies, setting it up, adding everything into a reaction pot and starting it stirring. He always loved that part of the experiment. He wouldn't find out until the next day whether the chemicals had reacted as planned or turned to brown sludge sometime in the night. Right now, as he stared at the gorgeous creature down in the kitchen, he knew the sky was the limit.

Will was the first to come into the house. "Good afternoon, Mr. Miller!"

"I still want your help tonight with that thing we talked about," Matt reminded.

"Jeb wants to learn, too," Will said quietly. "Jonathan has agreed to be the fourth. It will require some stealth if you desire to surprise my sister."

"The less she knows, the better," Matt replied, smiling.

Everyone had come into the dining room to sit as Mary put the food on the table. "We've something to do tonight," Will announced to the table. "We'll leave supper early."

"For what?" Grace asked.

"Mr. Miller, Jonathan, Jeb, and I have business in the hay barn," Will said.

"Business?" Grace said suspiciously.

"It's a private matter of Mr. Miller's," Will said. "I don't think it's polite to ask—"

"—a man his business," Grace finished for him. "I hardly think there'll be business in the barn."

"I won't break his confidence," Will replied. He looked hard at Jeb and Jonathan.

Grace turned to Jonathan, who shook his head with a resolute face. "A man's business is his own," the boy said. He would've sounded like a grown man except for his eleven-year-old vocal cords.

Thomas said, "Daughter, you would not expect the men to press you on matters concerning ladies."

"It feels not a private matter, the way they go on," Grace replied.

"A man's business is his own," Will said, adopting Jonathan's resolute expression. Everyone laughed.

"Keep your secret, then," Grace said. "It's obvious that some clandestine society has formed among you men."

Matt remained quiet, distracted by the prospects of dancing with this beautiful woman. He wouldn't be starting completely from zero; he had taken a semester of ballroom dancing in college to fulfill a physical education requirement. He hadn't done the dances more than a couple of times since, maybe at one or two fraternity brothers' weddings, and then sometimes to impress girls in clubs. It usually didn't take much, since other guys were usually horrible dancers. He faked it mostly, exaggerating the turns he knew and leading women around the floor until they became dizzy enough to pull close. He hoped that some of those skills would come back to him as they practiced.

When the dishes were done, Will motioned to the boys and they followed him to the barn. He had instructed everyone to bring a lamp. The barn was half full of hay, but there was still plenty of space to spread out. "Mr. Miller has much to learn tonight," Will proclaimed.

"Grace looks like a princess when she dances," said Jonathan.

"Great," Matt said sarcastically. "A princess?"

"Quite fortunate, actually," Will declared. "No one may be watching you at all." He turned to his little brother. "Jonathan, since Jeb and Matt will be expected to dance at the party, we'll be their training partners."

"We'll be the ladies," Jonathan said. "You might as well say it."

"Fine," Will said. "The ladies."

"Only if Jeb doesn't make fun of me," Jonathan replied.

"He won't," Will said. "But in any event, halfway through the instruction, Jeb can act the lady. Is this fine with you, Jeb?"

"Anything to dance with Sara," Jeb replied.

"There are three dances, mostly," Will announced. "The reel, the country dance, and the jig. If time remains, maybe we can practice the minuet."

"I had a class once where we did the minuet," Matt said. "I can't remember much."

"I'd not try it at a party unless you've mastered the steps," Will said. "But if you want to impress your mystery lady, that would be the way."

"Mr. Miller is trying to impress our sister, the mystery lady," Jonathan said. "That's funny."

"Naught is a secret on this farm," Will said. "Does anyone not know?"

"Probably the most important person," Matt said. "That should change in the morning."

"Compared to your trip tomorrow," Will said, "doing the minuet in front of the whole party may be easy."

Matt rolled his eyes.

It took them about an hour to master the reel. Many of the movements reminded Matt of the square dancing he did in college, and didn't take him long to develop some skills. "The most important thing about this dance is the eyes," Will declared. "As you move, you stare straight into her eyes like there's a string between your noses." He moved his hands, motioning back and forth between their faces. "If you do it right, your partner will melt to sweet honey." He pointed to his eyes while he was turning with Matt, and then watched Jeb do it with Jonathan. "Pretend he's Sara," Will said. "All the ladies want you to gaze at them like they are the most beautiful creature you've ever looked upon."

Will stopped them finally and said, "Now it's time to learn the jig."

"I like the jig," Jonathan replied. To Matt's amusement, the boy put his arms across his chest and did a rather competent jig.

"It would be easier with music," Will declared. He turned and left the barn with no explanation. While he was gone, the boys practiced what they knew about jigs. Matt found that the steps were easy and it was simply a matter of remembering who was supposed to go where. They stopped when they heard the barn door opening. Two more lanterns accompanied Will. "I brought the music," he said. His uncle and father followed him; David had a fiddle.

"I've not danced as of late," Thomas said. "May we join?"

"You can, Father," Jonathan declared, "but you should know that everyone has to take their turn being a lady."

His father laughed. "So no one can jest later?"

"Exactly," Jonathan replied.

"I'm here to play the fiddle," David said, laughing. "I'll not jest about anyone."

They spent the next hour learning the finer points of the jig from Will. His father was anything but light on his feet, so even though Thomas was familiar with the dance, his son spent more time correcting him than Matt. When they had finished the jig, the barn went quiet for a moment as Will organized them for English country dancing. During the silence, they heard a knock on the barn door. "Who is it?" Will called out.

"It's your mother. May I come in?"

Everyone looked around at each other and there was some mumbling.

"Are you dancing in there?"

"Maybe," Will replied.

"I'm coming in," she said. She opened the door with a lamp in hand. There were two other lamps behind her. Faith and Grace followed her into the barn.

Grace smiled at Matt. "If you should practice," she said, "you may as well practice with a real lady."

Jonathan looked around, counting. "One of us still must pretend," he reminded. "I've already taken my turn."

"I'll be a lady," Will said. "I'm a better dancer than most of them anyway."

"You are not!" Grace exclaimed.

"He is, dear," Mary said calmly. "Some of the ladies in the church don't dance well at all."

"Would it vex you if they join us, Mr. Miller?" Will asked, smiling.

It was obvious to Matt that the deal was already done. "Not at all," he said. "But they should remember that this is work, not play." They laughed at his deadpan pronouncement. The women set their lanterns down.

"We are teaching country dancing to Mr. Miller and Jeb," Will called. "Line up!"

David started to play the fiddle, and everyone fell into line. The women seemed already to know what to do, but Will would stop often to correct the men. Once they had mastered the country dance, Will switched from dance to dance to test their skills. The more they did this, the more comfortable Matt became with the movements, and after a while, he forgot himself in having fun rather than trying to perfect his technique. At one point, they all laughed like mad as they danced around the barn, kicking up a cloud of hay dust. Matt enjoyed watching Thomas dance with Mary, imagining them as a young couple. Matt's face ached from laughing and it made him realize that it had been a long time since he had really laughed, or even smiled.

It was two in the morning by the time everyone left to go back to their houses, leaving Matt with the dog in the barn. They never did practice the minuet, so Matt decided he'd try to win Grace's heart with the other dances. "I'll be damned if I didn't learn to woo Richmond women tonight, boy," Matt said.

The dog looked at him and tilted his head.

# CHAPTER 37.

## RICH MEN'S SONS

Matt was up with the sun. He washed and dressed in the clothes that Mary had loaned him on his first Sunday. Now that he had his new clothes, he realized that the borrowed clothing was not of the high quality he had first thought. The clothes were comfortable, though, so he thought he'd eventually ask Will if he could buy them. He stepped out of the barn after he was dressed and walked to the privy.

Thomas was already coming out. "I guess no one is required to wake you."

"I'm up at the crack of dawn like the other farmers," Matt replied.

"A few farmers are still sleeping after dancing late."

"I'd be sleeping too, if we weren't going into town."

"Maybe me as well," Thomas admitted. "Get some breakfast. Jeb will come. He wants new clothes for Sunday."

\*\*\*\*\*\*\*\*\*

Jeb answered the door chewing his food when Matt finally arrived at the house for breakfast. "Good morning, Mr. Miller," he said.

"I heard you're going into town."

"I desire new clothes for church."

"I thought they were for the Martins' party," Matt replied with a knowing grin.

"That too," Jeb said, yawning. "We stayed up late."

"I know, right?" Matt said, yawning back.

"I have some trepidation," Jeb said.

"It's all about the lady," Matt said. "Make sure everyone is looking at her."

Jeb stepped close to Matt to talk quietly. "Grace loves to flash it away when she's dancing," he said.

"Flash it away?" Matt questioned. "She shows off?"

"She loves the minuet," Jeb said. "Sometimes men don't ask, though."

"Why?" Matt said. "She's beautiful."

"Will said it's because she's beautiful that men fear to ask," Jeb replied. "He said you should still be the one to speak to the beautiful girl or ask her to dance."

"Your brother gives good advice."

*********

They started their trip to Richmond with Thomas riding at a relatively fast pace. There was little chance for conversation for the first mile, but then Thomas slowed Patriot and Matt found himself in line with the older man. It seemed as good a time as any to make his request.

"Mr. Taylor," he said. "I have something to ask."

"Sounds important," Thomas replied.

"It is." Matt felt one of the strings of light tugging him forward. When he saw them, they pulled him out of his current thoughts as he looked to where they led. He was silent until he could consciously bring himself back from his daydream. He had no idea how long he had remained silent.

"Well?" Thomas said. "Will you ask your question?"

"I want your permission to court Grace," Matt replied. He watched a huge grin cover Jeb's face. Thomas, on the other hand, had a strained look. It was obvious the request had taken him by surprise.

"You don't have the means," Thomas replied.

"Not yet," Matt said. "But I will."

"You will?" Thomas quipped. "The richest men in Virginia are asking for my daughter's hand."

"I'll be starting a business when I return to Philadelphia."

The older man laughed. "There's no guarantee of success," he said. "Then what?"

"Two years is all I need," Matt replied.

"Two years! She should already be starting a family."

"She's chosen not to," Matt said.

"She'd not have been around her animals."

"I'll not ask for her hand unless I can guarantee that."

"Every poor man within two days' ride would make the same pledge."

"I'm not a poor man. I have an education," Matt said. "If I have to come back with enough money to buy your farm, I will."

"Your word alone will not be enough to court my daughter," Thomas replied. "My answer is still no."

They rode for a while in silence, all three riders deep in their own thoughts, until Matt interrupted the quiet with a revelation. "Can I at least escort her to the Martins' party?"

Thomas shook his head at the boldness of this stranger who had somehow become intertwined with their lives. Eventually he said, "Only to the party."

They rode again in silence until Matt made the conscious effort to restart their conversation. "Jeb, are you going to Henry Duncan's?"

Jeb looked at his father, who answered for him. "No, Henry has little for young men." Thomas addressed his son next. "We should purchase your clothes before other errands." Jeb nodded. Thomas turned back to Matt. "You mentioned you had much to accomplish in town."

"I have a few places to visit and am hoping to spend time with Henry Duncan," Matt replied.

"Will you purchase more clothes?" Jeb asked.

"I told Henry I'd teach him my fighting style," Matt replied.

"Let's see how you fare with the horse in town, and then we can go our separate ways. We'll meet at two o'clock in front of Duncan's for the ride home."

"That would be good," Matt replied.

"It means you would lunch on your own," Thomas said. "It has naught to do with our previous transaction. I'm not trying to avoid you, now."

"The thought hadn't even occurred to me," Matt said, laughing. "I'll meet you at two o'clock. I'm not sure how long it will take to arrange things with Benjamin."

"Are you sure you desire intercourse with Benjamin?" Thomas asked.

"I need something to do until my ring sells," Matt replied. "My plan was to come into town and work with him."

"What about finishing your riding lessons?" Thomas said.

"I've asked to court your daughter," Matt replied. "I can't keep living in your barn."

"How long until you return to Philadelphia?" Thomas asked.

"My ring went to auction this week," Matt said. "I'll leave Richmond once it sells. One week…maybe two."

"I propose that you work for us another week," Thomas said. "The prospects of reaching a satisfactory agreement with Benjamin are dubious."

"One of my options would be to distribute my medicines through Benjamin's store," Matt replied.

"I believe you have the wrong impression of the man," Thomas said. "He's in some distress now because of his health, but even before this, his reputation was poor."

"I left him with a list of supplies," Matt replied. "He said he'd find them."

"Inform Benjamin that you are required to work another week and you'll leave for Philadelphia thereafter," Thomas said.

"Why would I do that?"

"The man's a lazy and witless soul," Thomas explained. "I'd not enter into any agreement with him. He'll be the first to yammer should you offer your medicine in Richmond. Tell him no more details of your affairs."

"I'm supposed to tell Benjamin I'm not interested?"

"If you learn that none of your items is available, then your decision should be made."

They were now entering the business district. Matt had been in the city enough now that he knew where he was going. "I'll meet you at Henry Duncan's at two o'clock."

"Any trepidation with the horse?" Thomas said.

"I'll be fine." He patted Thunder.

Matt headed to the silversmith. Jacob Berkley was sitting behind his bench with a lens in his eye, looking at a piece of jewelry when Matt entered his store. "How does it at the farm?" Berkley asked.

"Quite well," Matt answered. "We're done with the hay and the corn. Any word on the ring?"

"The first I'll hear is when the courier arrives this week, or the next," Berkley explained. "He'll have the gold and a bill of sale, or if it didn't sell, he'll have the ring."

"What if he does bring the ring back?"

"You must decide. I could buy it from you, or if you return the seven pounds, you can take it."

When they'd finished, Matt shook the man's hand, went out onto the street, and untied Thunder. "Let's go, boy," he said. He walked beside the horse as they made their way to the apothecary.

Benjamin Scott was sitting behind the counter, reading a book. "Can I help you, sir?"

"It's Matt Miller," Matt said, extending his hand.

"Oh, Mr. Miller," he replied. Still sitting, he reached his hand out to shake. "I apologize for not standing. My joints are aching more than is common this morning. I would buy more of that medicine from you, if you have it."

"There are only a few tablets left," Matt lied. "Could you get the supplies?"

"You didn't leave any money."

"You didn't say to leave money," Matt said coldly. He already knew where this was going.

"I'll buy some of that medicine," Scott said. "I much admired how it relieved my pain."

"I told you we could make all you needed."

"That won't help me *now*. I may never get some of those supplies."

Matt wasn't sure what was making him angrier, Scott acting like an idiot, or the fact that he hadn't seen the signs of this during their first meeting. "Which ones can't you get?" he asked, already surmising that Scott hadn't looked for any of the items.

"Everyone knows that you pay in advance."

"Mr. Scott, to let you know, I'm required to work on the Taylor farm for another week," Matt explained. "After that, I'll be returning to Philadelphia. It's probably fortunate that you didn't spend time looking for my supplies. You are relieved of any future obligation."

"I'm very disappointed that you've decided not to continue our business," Scott said. "It would have been beneficial to both of us."

Matt wanted to reach out and collar the man, but he forced himself to stay calm. "Have a good day, sir."

"What about that medicine?" Scott repeated as they shook hands.

Matt thought for a moment. He had come prepared with ten tablets wrapped in his pocket to barter for supplies. He'd need money if the ring didn't sell, so he thought of a high price. "Ten pounds for eight tablets."

"Ten pounds!" Scott exclaimed. "No medicine's worth that."

"It's very rare, and you'd have to go all the way to China for more." He had little sympathy for Benjamin Scott, so turned to leave. "Good day, Mr. Scott."

"Eight pounds for eight tablets," Scott said as Matt reached for the door.

"They're worth more," Matt said, turning back. "Plus, there are the four I already gave you."

\*\*\*\*\*\*\*\*\*\*

Matt walked out of the apothecary shop with nine pounds and eight fewer tablets of the world's supply of ibuprofen. It would be enough money to purchase the ring back if it didn't sell and he was exceedingly glad to be rid of Benjamin Scott.

Matt rode Thunder for the few blocks it took to get to Henry's store and tied him outside. Henry was finishing with a customer. "I'll be with you in a moment, Mr. Miller," he said. Matt wandered around looking at the racks of clothes. Much of it was elaborate and not his style, but he could imagine himself in some of the simpler items. "Here for more clothes?" Henry asked once his customer left.

"Not unless you've something for a party," Matt replied.

"You're speaking of the Martins'."

"That's the one," Matt said. He pulled a decorated jacket from the shelf, scowled and put it back. "I don't like the fancy stuff."

"You liked your clothes last week?" Henry asked.

"Best-dressed man in Richmond."

"Did you make a favorable impression on Miss Grace?"

"I've been given permission to escort her to the Martins' party."

"Mr. Miller," Henry exclaimed. "Congratulations!"

"Hardly congratulations," Matt replied. "I asked to court her and her father said no."

"Still 'tis a laudable accomplishment," Henry said. "God has some plan for you." Then he chuckled and said with some irony, "Of course, I can imagine with the beauty of Miss Grace, more than one man in Richmond believed 'twas God's plan that she be his wife."

"I have to believe," Matt said, laughing.

"What brings you to my shop this fine day?" Henry asked.

"I thought I could show you my fighting style," Matt said. "I'll go eat lunch and come back when you're free."

Henry stepped to the door and turned the sign around to read "CLOSED." "We should do it now," he said. "There are victuals in back and a table outside where we can watch people pass by." Henry motioned for Matt to follow him to the back of the building. They worked together to move plates and food out to a wooden table and soon sat in front of meat, bread, wine, and cheese.

Henry said, "Tell me how you met the Taylors," and Matt spent most of the meal explaining.

"And so after all this," Henry said after he was done, "you don't know how you got under that bridge?"

"I don't," Matt replied. It was as truthful as he thought he could be. He wasn't sure there was anyone in the colonies who could handle a story about time travel.

"You already have enough money to hire passage to the North," Henry said. "Why haven't you gone?"

"The obvious reason is Grace," Matt replied.

"There are many beautiful ladies in the colonies."

"The whole lifestyle's attracted me."

"Richmond couldn't be much better than Philadelphia."

"Traditions and expectations," Matt replied simply.

"What expectations?" Henry asked.

"Where I come from, people never say a prayer before they eat," Matt explained. "They don't get dressed up and go to church. We live our lives in Philadelphia without much regard to the others in the community."

"So it's church and community?"

"Yes—no. I was starting to feel lost in Philadelphia."

Henry shook his head, smiling, but still stared at Matt, waiting for more.

"You have to get permission from a woman's family to court her here," Matt explained. "I never met my old girl-friend's parents."

"Family's important in Virginia," Henry said.

"There's a culture of honor and responsibility that I'm not used to. It's challenging, but every time I manage to meet their challenge, I feel exhilarated."

"I understand. Though 'tis not everyone in Richmond. Thomas has labored many years for what you see. The Taylors are endearing because they don't own to their prominence."

"I don't think of them that way," Matt said. "The Martins, on the other hand..."

"They do have a bit of the plum about them," Henry said. "Speaking of, you've taken quite a responsibility escorting a Richmond gentlewoman to a society party."

"Society party?" Matt said. "It's a picnic."

"My boy!" Henry exclaimed. "Richmond takes its picnics very seriously. How's your dancing?"

"I practiced last night. I'm ready, except for the minuet."

"You've never done the minuet?"

"I had a dance class. It's been a long time."

"I know our charge these next hours."

"Practice fighting skills, I thought."

"Fighting can wait," Henry replied. "You're speaking to the best dance instructor in Virginia."

"How did I not already know this?" Matt said. "You've time to practice dancing?"

"I can close the shop."

"I thought I'd skip the minuet."

"What about meeting their challenge?" Henry exclaimed. "None of it seems rational, but if you should lead Grace onto the floor and show her properly to Richmond, she'll consider you the angel Michael himself. You should only incur the cost of a new cravat."

"What's wrong with the one I have?"

"You'll not want to wear that old thing."

"I just bought it!"

"You said naught of a party," Henry replied.

After lunch, they spent a full three hours dancing the minuet. Henry had a studio in the back of his building. He pushed fighting targets out of the way to make an area for dancing. He'd count out the beats as he moved Matt through the dance.

After about an hour, Matt said, "I have it, Henry."

"Not even close!"

After the second hour, Matt asked, "Am I getting there?"

"Closer," Henry said. "You still hesitate."

"It's hard to remember all the forms."

"Grace has been doing them since she was a girl," Henry warned.

"I thought everyone would be looking at her."

"Not likely. Keep dancing."

Matt's legs ached by the time they heard knocking on the door. "Your escort home," Henry announced. Matt followed him to the front of the store. It was Thomas and Jeb.

"A few more minutes," Henry told them as they walked in. "Let me get that cravat for you, Mr. Miller. It's ten shillings."

"Ten shillings for a cravat?" Jeb said, surprised. His father gave him a dirty look.

"Let that be a lesson to you, my young man," Henry said. "Cloth of the finest quality is never too heavy. Don't you agree, Mr. Miller?"

"If that means a good necktie is expensive," Matt said, "then I agree." It was true even in his own time. He had always marveled how such a thin strand of silk could cost so much. He fished coins out of his pocket and exchanged them for the cravat. "Grace ties them," Matt said.

"Then she appreciates a good silk," Henry said knowingly. He reached up on the shelf. "This is a diagram to that fighting style we practiced. I'd step though it a few more times tonight so you can commit it to memory."

Matt opened the pages. It was a diagram showing the steps for the minuet.

"Thanks, Mr. Duncan," Matt said. "I'll practice."

"Mind that you do. You never know when you may want those skills. Good day, gentlemen." Henry shook everyone's hand.

"Will you be at the Martins'?" Matt asked Henry.

"There's a rumor that there'll be dancing," Henry said. "Henrietta and I wouldn't miss it."

"Henrietta?"

Henry only smiled. "Good day, gentlemen," he repeated.

# CHAPTER 38.

# FIVE MINUTES

---

Matt was tired on the ride to church. He had practiced the minuet in the hay barn late into the night. There were a few baroque composers among the albums on his phone, and these seemed to have the proper tempo for the dance. He could see the minuet patterns in his head after looking at the diagrams so many times. Anyone coming into the barn would've thought him mad as he bowed to and danced with his imaginary partner. Towards the end of the night, he had focused so hard that he could practically see, feel, and hear Grace as they danced together in his mind.

At times, it felt bizarre that so much depended on dancing. He had asked himself more than once whether learning to dance was really worth his time. He sat a while in the barn during a break to answer this question in his mind and had decided that the best answer was that "it seemed like it." In the big picture, it mattered as much as anything else. It was no less important than many of the other seemingly trivial things in life, which included getting to work on time, respecting elders, and not cheating on your girlfriend. Your life wasn't over if you didn't do them, but you usually had a much better day if you did.

Grace had tied his cravat for him in the morning and commented on how beautiful it was. It started him wondering whether cravats mattered in the scheme of things. Again, the answer was that it seemed like it; especially when a beautiful woman had to stand excruciatingly close to tie it properly. He remembered the smell of her hair as she hovered around him and the feel of her ribbons as they brushed his face. He knew she was looking forward to the party. He, on the other hand, was nervous; he had no idea what to expect. They would go after church to eat, dance, and socialize. The more he heard about the party, the more overwhelming it seemed.

Jeb sat in front of him in the wagon, wearing his new clothes. The teen wasn't used to wearing a tie, and he shifted often on the bench and twisted his neck. Mary and Grace had new dresses and Matt wondered if they were part of the "Martin Collection." The cloth was of obviously high quality. Mary's dress was royal blue and Grace wore white with red trim. Thomas had dressed in a dark blue suit that wasn't new but still made him look sharp. He carried himself with a formal but comfortable demeanor.

Will followed them in his buggy since he'd be staying in town. David and Faith would come later, once Charles arrived to watch the farm. Will wore clothing that was more colorful than his father's. He had been in an exceptionally good mood all morning. Jonathan was the only one who seemed truly uncomfortable. He fidgeted beside Matt, tugging at his clothes and voicing multiple complaints, including that his pants were itchy, his collar was too tight, and the jacket was hot.

Grace turned around. "Are you ready to dance, Mr. Miller?"

"I won't embarrass you, if that's what you're worried about," he replied. It felt strangely awkward now that his attraction to Grace was out in the open.

"You're a capable dancer," Grace said, smiling.

"The only one I'm not comfortable with is the minuet," Matt replied.

"We can put ourselves far down on the list," Grace said with disappointment in her voice, "as long as we dance sometime."

"There's a list?"

"Everyone signs," Grace said. "The hosts select couples for the honors."

"Honors?"

"It's usually the first five. After that, couples dance in order of the list."

"How does the host select the top five?"

"The first dance is always the hosts," Thomas said. "After that, it's whoever has cause. Ofttimes, honors go to prominent couples making some statement; other times to the best dancers. Honors can be a knee-knocker with everyone looking on."

"It's brilliant watching the first five," Grace proclaimed. "We already know one who claims honors."

"Who?" Matt said.

"My brother," Grace answered, irritated. "He's asked Graine if they could dance second."

"Daughter, you'll have many opportunities to dance honors," her mother replied.

*********

You could have asked Matt about church right after it ended, and he'd not have remembered anything. He spent the entire service dancing the minuet in his head. One time he had to recover from putting his hand on his belly and bowing in the church pew. Grace caught him, so he pretended that he was leaning forward to fix his coat, and then shuffled around to emphasize the ruse.

When they pulled up to the Martins' house, servants came to take the wagon and Matt heard them commenting on the beautiful horses. Graine and Elizabeth Martin greeted the guests as they arrived. The Reverend Michael had given the Martin women special dispensation to miss worship on this Sabbath to prepare the "celebration of God" at their house. It seemed easy to get a dispensation from the Reverend Michael as long as you threw a good party and he was invited.

Matt saw Robert Martin arrive from church and park his wagon at the side of the house before hurrying over to step to the welcome line in front of their house. Matt waited with the Taylors, standing beside Will to enter the party after Grace, her parents, and the younger boys. He looked down the line and realized that there were more than a few women doing the greetings.

"Who are all these women?" he whispered to Will.

"The Martin daughters," Will replied. "Graine has six sisters."

"Guess he's given up on a son," Matt said.

"Most likely," Will replied. "He's anxious for the first to be married so the others can follow." Graine was the last sister in the line. She stood there with a brilliant smile on her face. The girls were lined up according to their age and got progressively taller down the line. Graine was almost as tall as her father.

"Too bad the eldest sister is so homely," Matt said. "The others may never marry."

"Homely?" Will asked. He looked down the line for a moment until realizing that Matt was joking. "Oh, she does look beautiful." As they reached the receiving line, Matt made sure to let his friend lead so he could see how Will treated each family member. Matt talked to Robert more than he had expected. He asked the man how he was doing and it became a lively discussion about preparing for today's event and dealing with the Martin women. Robert Martin and his wife Elizabeth were a friendly and dynamic couple and their influence on Graine was evident.

Matt watched Will carefully when he came to the daughters. Most blushed as he gently kissed their hands and said hello. Will was a charming man and the young girls put giant smiles on their faces as he greeted each of them in turn. Matt delayed with most, asking them their names and ages and doing a lot of hand kissing. He spent extra time with the younger sisters, who were giggly and fun. He was able to forget the minuet briefly, but as he neared the end of the line he became impatient to talk to Graine, who was being held up by Will. When Matt finally reached her, he offered a quick hello, pulled her aside, and whispered into her ear, "Can I ask you for a favor?"

"Certainly, Mr. Miller," she said, surprised.

"I don't want Grace to hear. Can we have one of the honor positions for the minuet?"

"You and Grace? Those are usually for married or courting couples."

"I'm her escort today."

"Congratulations!"

"Can you get us the position?"

"There are others who belong there."

"Grace assumes I'm going to sign up later and she's fine with it," Matt said. "I want to surprise her."

"Oh!" Graine exclaimed. "That does sound romantic."

"I return to Philadelphia soon," Matt said. "I need to leave a lasting impression." He could see that the family behind him was getting impatient.

"Mr. Miller, I can't make this pledge. I must speak to my parents."

"Thanks, Graine," Matt said. "I'll be a happy man if I can have that dance." He reached down and grabbed her hand, then looked into her eyes. "Even if you can't, let me say that you look ravishingly beautiful today." He kissed her hand.

"You've been spending too much time with Will," Graine said. "You're becoming hopelessly charming."

Matt turned and stepped away towards Grace, who had been waiting for him off to the side.

"What did you say to Graine?" she asked.

"She was mentioning how beautiful you look in that dress," Matt replied. Grace had her hair arranged in braids that wrapped around her head. It was a big change from her usual no-nonsense, pulled-back hairstyle. Like most naturally attractive women, Grace would've looked good wearing a sack, but when dressed up, she was a breathtaking beauty.

"Are you already flirting with other ladies?"

"With Athena herself on my arm," Matt said, "I hardly think that would be wise."

By her expression, Matt knew he had said the right thing. He held out his elbow, Grace put her arm through it, and they walked into the crowd of people.

"All joking aside," Matt said, "I don't know what it means to escort someone to a Richmond party, besides walking around and smiling."

"You've never taken a lady to a party?"

"If I wanted to make you look exquisite at this party," Matt said, "what would you have me do?"

"I'll introduce you," Grace said. "You ask questions and make your acquaintance. You're allowed to flirt, but the ladies should know you're my escort."

"So now I'm allowed to flirt?"

"It shouldn't be hard. You're a handsome man with a strange accent."

"And what will you be doing?"

"Standing beside you, also chatting, and acting unimpressed by your adoring stares."

"Shouldn't be hard for you, either," Matt replied. She pulled her arm from his and reached down to grab his hand with warm fingers, and they walked together into the crowd that was steadily growing in the yard around the Martins' home.

Grace introduced him to many people and he found that he had no problem engaging them in conversation. He had no idea how people lived in the colonial South and was enthralled with even trivial details, and it was easy to get people talking about their lives. One man gave Matt a crash course on the tobacco business, and after twenty minutes, shook his hand and said, "Thank you, Mr. Miller, you're one of the most interesting men I have ever met." Matt smiled at the irony. He had said maybe ten words, and all were questions about the man's expertise.

Matt met Grace's friends and tried his best to flirt, but it wasn't hard to return his attentions to Grace when appropriate. With this beautiful woman at his side, he felt more handsome and charming than ever. He stepped back on more than one occasion and reminded himself to take stock of what was going on. He wanted to remember this experience. They mingled until a bell rang. Matt felt a splash of adrenaline surge through his body as he thought that it might be time already for the minuet.

"The victuals are out," Grace said. She hooked her arm through his and led him toward the line of people moving towards the serving table. She pulled him in between what looked like a guesthouse and some servants' quarters. "I thought I'd never get you to myself," she said before she kissed him hard. When she finished, she lingered at his mouth, brushing her lips against his. Matt looked around briefly to make sure no one was watching, suspecting that this type of activity was frowned upon. She leaned in and kissed him again. It was less forceful this time and had a silky finesse that tickled his mouth and remained on his lips. She stepped back, smoothed her dress, and said, "Are you ready to eat?"

It took him a moment to find his words. "I think so." He looked around, again. "You're not going to get me into trouble, are you?"

"I'll no longer devise ways to for us to be alone if it causes you such trepidation," Grace said.

He smiled. "I wouldn't say trepidation."

"Good." She leaned up again for another long kiss and then pulled him back towards the crowd.

The feast centered on a roast pig colorfully decorated with cherries, apples, and pineapple. Matt wondered how far they'd had to go from Virginia to get a pineapple. Servants in livery ladled food onto their plates as they walked down the banquet line. When their plates were filled, Grace led them to her parents who sat at one of the long wooden tables that had been set up under the great oak trees scattered about the estate. Thomas had a big smile on his face. "Have a seat, young people," he said.

"Enjoying yourselves?" Matt asked as he and Grace took their seats.

"I am," Thomas said, looking at Mary. "My wife looks exquisite this day."

"You enchanting man," she said. She laughed like a young girl.

"You do look very pretty, Mrs. Taylor," Matt said. "It's easy to see where your children get their fine looks."

"You've been spending too much time with my son, Mr. Miller," she said, "with all these compliments." She smiled and then looked at Grace. "Are you having fun, dear?"

"Aside from the fact that Mr. Miller has proven himself to be very shy," Grace replied.

"Why, Mr. Miller," Mary said, "I'd not have predicted this of you."

"She's joking," Matt said. "Grace had to drag me away."

"I do jest," Grace said. "Mr. Miller talked to Mr. Connell about tobacco forever." She rolled her eyes.

"You know the tobacco trade, Mr. Miller?" Thomas asked.

"No," Matt replied, "but Mr. Connell was polite enough to tell me."

Grace interrupted, looking at Matt. "The dancing will begin soon," she said. "We should sign."

"I already asked Graine to put us on the list," Matt replied. "It's lower so we can watch plenty of other couples first."

"This is why you spoke to her?"

"I didn't want to end up in an honor position accidentally," Matt said. He could see the disappointment in her face.

"As long as we dance," Grace said, "I'll be pleased."

"I'll make a point of practicing for our next party."

Grace smiled back at him and then scanned the crowd. "Where are Jeb and Jonathan?"

"Jonathan has found boys his own age and they have removed most of their clothes," Mary replied. "Jeb is nearby."

"Jeb looked handsome in his new clothes," Grace proclaimed.

"Sara and her friends came by to ask him to join in their party games," Mary said.

Their conversation was interrupted by another bell. "The minuet is starting!" Grace exclaimed. "We should find our place."

"I'm not done eating yet," Matt said. They hadn't had breakfast and he was hungry.

"Please hurry," Grace said. "I'm anxious to see who the Martins have selected."

"How long before it starts?"

"Fifteen minutes from the bell."

"I get five minutes to finish my food," Matt declared. If by chance he did have to dance in front of the entire population of Richmond, he thought he should be able to eat first.

"Fine," Grace said. "Five minutes."

# CHAPTER 39.

# MINUET

---

Matt wolfed down as much of his food as he could before Grace pulled him to his feet, saying, "Let's go."

Her mother gave her a disapproving look. "Grace, dear, where are your manners?"

Grace rolled her eyes and pulled Matt to the dancing pavilion. It was a large open-roofed building with a polished metal-grey cobblestone floor, about the same size as the pavilion Matt had stacked full of hay. It allowed plenty of light, but protected the dancers from the sun. It seemed that every young woman in attendance already had the idea of staking out a place to watch the dancing. Many of the young men stood tentatively behind their dates, looking like they would rather be somewhere else. Violins, flutes, and fiddles already played, and Matt recognized the three-quarter time signature that was the hallmark of the minuet.

Matt began to feel nervous again, but pushed it to the corner of his mind after reassuring himself that he had spent almost six hours practicing the dance. He felt like a prizefighter getting ready for a fight. An announcer from the band stepped out, bowed to the audience, and spoke loudly. "Welcome, friends, to the home of Robert and Elizabeth Martin." He motioned to Robert and Elizabeth, who waved from their position next to the band.

Robert exclaimed loudly, "Welcome, everyone!" The crowd applauded and broke out into a cheer.

The announcer spoke again. "I'll be announcing the honor portion of the dance. Selections for the top five positions were made only moments ago from those who signed this day. The couples' names will remain secret until they are called."

Grace smiled cheerfully and turned to Matt. "It's so much more suspenseful that way," she said.

"Yes," Matt replied. "That does seem suspenseful." He didn't want to ruin her fun, so he tried to sound enthusiastic. Grace wasn't paying attention to him anyway, so he thought it might not matter one way or another. The announcer stepped off the floor and the musicians started playing again. This went on for what seemed like an eternity. Matt imagined it had something to do with building the suspense. After a while, the announcer stepped out onto the stone floor and motioned to the musicians to lower the music.

"The first honors will be our hosts, Elizabeth and Robert Martin," he said. The crowd cheered as they walked out onto the floor. As the music continued to play low, the couple moved in unison to face all sides of the pavilion to recognize their audience. Once this was complete, they came together, stepped back, and bowed to one another.

Matt leaned down to Grace and said in her ear, "Does everyone do the dance the same way?"

"Yes," she said. "Everyone does the same forms." This was good news. Matt would be able to watch the couples and get an idea of spacing and patterns on the stone dance floor. When the music started again, the dance began. The Martins seemed like competent dancers, so Matt focused like a laser on the couple as they moved, watching their feet and spacing. Henry's diagram and instruction had been dead-on, and Matt was able to superimpose their dance on top of the diagram in his head. They bowed to each other when they'd finished and then stood side by side, holding hands.

The announcer walked out beside the hosts and waved his hand toward them. "Ladies and gentlemen, I present to you Elizabeth and Robert Martin." The crowd cheered, and the couple waved and stepped off the floor.

"A very competent demonstration," Grace said. "Both have come far under Henry."

"Henry Duncan?" Matt asked.

"Henry is a magnificent dancer and does lessons."

"You don't say?"

She looked at him suspiciously. "What are you up to?"

"Nothing," Matt said, trying to look as innocent as possible. "I thought the only thing Henry knew was sword fighting."

She didn't reply and still looked at him suspiciously, trying to read his face. Her scrutiny was interrupted as the announcer again stepped out onto the floor. "My friends, the next dancers are…Miss Graine Martin and her escort, Mr. William Taylor."

"I told you he'd get honors," Grace said. "He'll be impossible after this."

"Not if he messes up," Matt said.

"He won't, and even if he does, he still manages to make them laugh."

"When we're out there," Matt said, "I plan to do funny things, too."

"Thankfully most of the crowd will be gone," Grace said. "If you don't take this seriously, I will be angry with you for the remainder of the evening."

"I'll take it seriously," Matt said, "even if no one is watching."

She turned away as Graine and Will walked out onto the floor. They bowed to the audience and to each other. Matt focused again, visualizing the steps. He put himself in Will's place and danced with Graine in his mind. There were only two times where he moved incorrectly in his imagination. The dancers ended and held hands, bowing to the audience.

"Ladies and gentlemen," the announcer said beside them, "Miss Graine Martin and Mr. William Taylor." The crowd applauded.

He caught Grace staring at him when the dance was done. "What?"

"I have never seen a man so focused on the minuet."

"You must admit, they are a very attractive couple," Matt said.

"We would be as attractive."

"You'll not get any disagreement from me. Next time."

The announcer stepped out on the floor again. The music was playing quietly. "The next couple will be..." He paused for a long time while the music played to build the anticipation. "Mr. Henry Duncan and Miss Henrietta Mordeau."

"I knew it," Grace said. "Henrietta is in town."

Matt was dumbfounded. "Isn't Henry a confirmed bachelor?"

"She comes from Europe twice a year," Grace whispered up into his ear. "She's an exquisite dancer. They deserve the honors." Henry and Henrietta stepped out on the dance floor. Henry wore a powdered wig and was opulently dressed and his partner was equally decked out.

"The French king and queen," Matt proclaimed. *That's almost too funny, Henry and Henrietta.*

"Quite elegant," Grace said. "Miss Mordeau is almost heavenly in her grace and beauty."

Matt focused on their movements as they danced. They *were* very graceful and brought a new perspective to the dance, but fortunately for Matt, they too followed the same form. He was able to imagine it completely in his mind this third time with no missteps. The crowd roared with applause when they finished.

"They never disappoint," Grace said.

"I know I'm not disappointed."

She slapped him lightly on the arm. "We'll see someday when we have honors how well you do."

"Careful, Grace," Matt warned as he smiled. "I may raise my hand and volunteer us right now."

"You would not."

"I could make it up as I go."

She shook her head and pointed. They were ready to announce the fourth couple. Matt was starting to have the feeling they wouldn't be called, so he relaxed. The announcer stepped out again. "The fourth couple to hold honors for the afternoon is...Miss Annabel Creighton and her escort, Mr. James Carlson."

"There's a rumor that they will be engaged soon," Grace whispered. "Many suspected they would be chosen. They aren't the most capable dancers, but it's important to make this statement."

"A lot depends on this minuet," Matt said. He wasn't entirely joking. This felt more stressful than necessary.

"A gentleman must manifest his commitment to the lady he loves," Grace said.

"I get it," Matt replied.

A woman at Matt's side tapped him on the shoulder and put her finger to her lips. Matt whispered an apology and focused on the dance. As Grace had said, the Creighton girl and her fiancé did the dance competently, but both looked mechanical. It was to their disadvantage too that they went immediately after Henry and Henrietta, who were minuet royalty. When they finished, the crowd, aware of their challenge, clapped as loudly as they had for the others. Both man and woman smiled as they walked off the floor. Matt had to laugh when the young man gave an audible sigh as he walked past.

"Last one," Grace said. "'Twill be Gregory Smith and Anne Marie Moore. They are next to be married."

"We'll see," Matt replied. He put his finger to his lips. There was some nervousness in his knees, and he consciously had to steady himself. The music went quiet, and it took a while for the announcer to walk out onto the floor. He was shameless in the way that he played the crowd, building anticipation. Eventually, he stepped forward.

"Now, ladies and gentlemen, we come to our last honors position." He paused. "Who will it be?"

Matt could feel his heart pounding. *Oh, my Lord, you idiot, get on with it!*

"The last honors position goes to…goes to…Miss Grace Taylor and her escort, Mr. Matthew Miller."

Matt tilted his head to the sky and mouthed the words, "I need you now, big guy." He gazed expectantly at Grace and motioned for her to take his hand. She looked at him with horror and surprise, and it took every ounce of courage for Matt to take that first step and lead her into the square. The music continued to play low as they moved onto the floor. Surprisingly, once he was out there and separated from the cheering masses, a wave of calm and confidence swept over him.

"You don't know how to do this," Grace whispered loudly. "We shall never live this down."

"You better do your best to show me the dance, then," he replied.

"This is how you'll take your revenge, is it not?" They now bowed to each side of the pavilion.

"Revenge for what?"

"My animosity toward you those first days."

"You admit it!"

"I apologize. Can we step off?"

"Apology accepted." They bowed to one another and took their positions. "Too late, though." The music started to play and he began to move through the steps, one after another. Both started tentatively, but as they danced, the link between them grew stronger. At times, Matt felt almost as if he was watching himself dance from outside his body somewhere up on the ceiling. *Make her the center of attention.*

As they moved around the square that dictated the form and movements of the minuet, he made sure always to focus on his partner. She was stunning in her white dress and braids. When she moved, he matched her exactly and would follow by flipping his hand towards her as if to say, "Look at this beautiful woman." Then, almost as if it had happened in an instant, the dance was over. For his life, Matt couldn't remember if he had done it at all, or if he had skipped half the steps. As they stood there bowing, the announcer moved out to them, waving his hand, and said, "Our last honors dancers, Miss Grace Taylor and Mr. Matthew Miller." Grace slid her hand into his; he could see tears in her eyes. He led her slowly out of the square.

As they stepped off the stone surface, Matt took her possessively in his arms and kissed her passionately, not caring who saw. Grace responded in kind, pressing her lips hard against his. When the kiss was finished, she pulled away and stared into his eyes with a look of desire, then turned and walked away. Matt stood there, over-whelmed.

"Why, Mr. Miller," said Graine from behind, "I wouldn't have believed that you could perform such a competent minuet. Grace told me earlier that you weren't interested in dancing."

Matt turned to face her, coming out of his trance. "You took a big chance, then?"

"I imagined," Graine said, "until I spoke with Henry and he said that he taught you. I'll be making apologies to Anne Marie Moore until I can rectify the situation at our next party."

"Sorry about that," Matt said. "I do appreciate the opportunity. Did I make an impression?"

"Too mechanical in your turns, was Henry's comment," Graine said, "but many young ladies sighed when you did that motion with your hand. Were you casting a line to your young love?"

*Casting a line to my young love? What?*

Will stepped beside Graine and she casually put her hand into his. "No matter what your sisters are saying," he said, "ours was better."

"It's true that our dance was technically better than theirs," she replied, "but Mr. Miller made many more young ladies sigh with his adoration for his partner."

Will shook his head in mock disgust at Matt.

"I must go help Mother with the desserts," Graine said. She smiled at Will.

Matt watched her walk away. "You're a lucky man," he said to Will.

"She's planned endlessly to ensure everyone leaves happy," Will replied. "Anyway, how do you explain all that?"

"All what?"

"We never practiced the minuet."

"I went to visit Henry Duncan," Matt said. "We spent three hours practicing."

"You spent three hours practicing the minuet?"

"More like six," Matt said. "Remember how I left early to go to bed? I went to the barn to practice from diagrams he gave me."

"Who taught you that hand motion?" Will waved his hand, trying to reproduce it. "Something about it made them all swoon. Did you kiss her at the end?"

They were interrupted by a man who had walked up to them quickly. "Will," he said angrily. "Where's your father?"

Will looked over his shoulder and said, "Good day, Edward. Have you met Matthew Miller?"

"I don't care to make anyone's acquaintance," the man said. "Where's your father?" It was then that Matt saw Jeb behind the man.

"Hello, Jeb," Matt said. "Having fun?" Jeb was silent. He motioned with his head at the man standing in front of him.

"I caught him in the barn with my fourteen-year-old daughter," the man said.

"Is this true, Jeb?" Will asked.

Jeb mumbled to the ground. "We only kissed a couple of times."

"I don't want details!" the man exclaimed.

Matt was sure they were in the presence of Mr. Edward Greene, father of Sara.

"Come hither," Will said, motioning for them to follow. He turned to Matt. "Probably best that you don't come." Matt watched them walk away with Jeb following about five steps behind. He turned around once and Matt shook his head, smiling. He caught himself, thinking it might not be appropriate. Matt wasn't sure exactly how much trouble Jeb was in, so he decided he'd not make light of it until he figured out how bad things really were.

He looked around for someone he knew and saw only Grace, who was having an animated discussion with a group of young women. He stood there watching couples minuet until he noticed Henry Duncan on the other side of the pavilion. He put his hand up to get his attention and Henry waved him over.

"I can't thank you enough, Henry," Matt said, walking up to shake his hand.

"I didn't think you'd vie for an honor position," Henry said.

"What'd you think?" Matt asked. He realized he was fishing for compliments, but didn't care.

"Your technique was loose in almost every regard, and you have a long road to becoming competent—"

"That bad?"

"If you'd only let me finish," Henry said. "You made up for your lack of technique with desire for your partner. Even my Henrietta was mesmerized with your adoration for young Grace. I have rarely seen Henrietta react so, and she's danced with the finest practitioners in Europe. She said that every time you waved your hand, it was as if you caressed your partner from afar. I felt that you were pointing and trying to divert attention, but that's my guess and you needn't disclose your true intent. Sometimes the passion in the dance overwhelms all watching. Technique or no, this was that day for you, Mr. Miller. Now, to end my pontificating and answer your question, you did more than fine today. Young Miss Grace will remember this day for the rest of her long life."

"Thank you, Henry," Matt said, "even if you're just being polite."

"Whether I am or not," Henry said, "you won't always be able to count on mesmerizing young ladies with a wave of your hand. When you return to Richmond, I'll expect you to contact me with a formal request for dancing lessons."

"I'll be back," Matt said. "Anyway, who is this Henrietta?"

"A lady of exquisite breeding," Henry explained, "who comes to honor me with her presence a couple times a year."

"Someone told me that you were a confirmed bachelor."

"A valid assessment. Henrietta knows this about me, but she still visits. She's a wealthy widow. I cherish her presence and dread the day when she finally tires of my bachelorhood."

"I wish you the best future with your friend. Do you visit her?"

"Twice a year," Henry said. "I travel to Europe and spend time at her estate." He chuckled. "She feared once that I coveted her money, but knows differently now."

"How long has this been going on?"

"Ten years."

"Where I come from," Matt explained, "it's rare for men and women to stay married for ten years."

"I never take her for granted. I know there's pressure for her to remarry."

"And yet it's been a decade," Matt said. "I'm envious of your friendship. She's a striking woman."

"I agree," Henry replied. "She's a better dancer than me and magnificent with a sword."

"You're joking, right?"

"No, 'tis the truth," Henry said. "She's employed a sword master on her estate since we met."

"You're an interesting guy," Matt replied. "Would you mind if I wrote to you when I return to Philadelphia? I think you would give me the clearest picture of the happenings in Richmond."

"Matthew Miller," Henry said. "Are you asking me to spy on the Taylors?"

"Nothing of the sort. Leaving Richmond may make me homesick."

"You may write to me. Be aware that if I've not written back, fashion has called me across the sea."

"Thanks," Matt said.

"The dancing is starting again," Henry proclaimed. "I must find my Henrietta. My time with her is precious."

"I should join Grace," Matt said, "but she seems so happy talking to her friends."

"She'd appreciate being rescued after this much time, I believe."

"See you on the dance floor," Matt replied.

"Henrietta despises the reel," Henry declared. "I make her do it anyway." He laughed a hearty laugh, turned, and walked away.

Matt stood there, enjoying his solitude. He had done the dance.

# CHAPTER 40.

## IT WAS HER IDEA

"I have never been so embarrassed," Mary said on the way home.

"'Twas not my conceit to enter the barn," Jeb said. "Sara suggested it."

"The ladies love a man with—"

"Jonathan!" Matt interrupted before he could finish. Thomas glared at them both and then focused on Jeb.

"Why didn't you speak thus when Mr. Greene told the details?" his father asked.

"I didn't want her relations to think ill of her," Jeb said. "I'll take any punishment." He looked off in a dreamy fashion.

His tone wasn't lost on his father. "Since you didn't respect Mr. Greene or his daughter, it will be a very long time before he grants you an audience, if ever," he said. "You must consider the consequences before you commit such indiscretions."

"I'll not be allowed to speak with her?" Jeb asked.

"If it were one of my daughters," Thomas said, "it would be a very long time."

"Protecting her fame," Grace said. "That showed character."

"There should be naught positive in this!" her mother replied.

Grace shrugged. "A lady remembers," she said. Matt sat there silent, smiling. He couldn't add any wisdom to an already confused situation. Jeb had that far-off look in his eyes and Matt suspected he'd do it all over again.

"Fortunately, this matter was handled discreetly," Thomas said. "Jeb will receive the appropriate punishment from Mr. Greene. Otherwise, it was a wonderful party with more than a few surprises."

Mary said, "My son was such a gentleman to Graine, and their minuet captured everyone's imagination." She went silent and it was obvious she was hiding her smile.

"You should not tease your daughter thus," Thomas said, laughing. "Tell her."

"'Twill only grow her conceit," Mary said, "like your eldest son."

"Wife," Thomas proclaimed, "you know I'll not say it as well."

"Fine," Mary replied with a big smile. "First, who knew Mr. Miller could minuet, and under the pressure of honors?"

"You danced honors?" Jeb asked, surprised.

"You may have been busy during that time," Matt said.

"Oh." Jeb went quiet.

"It was an excellent surprise when they called your names," Mary said. "Your father choked on his tea."

"Not choked!" he said. "It was only unexpected."

Grace said, "Father, I didn't know you cared so much for the minuet."

"I knew what dancing honors meant to you." Thomas paused, thinking. "And I must apologize to Mr. Miller. I worried that my daughter would be embarrassed. I shouldn't have assumed this."

"The spot wasn't guaranteed until our names were called," Matt said. "I was as surprised as anyone."

"I answered many questions this day about the tall mystery man my daughter had chosen as her escort," Mary said. "Such a demonstration. The young ladies dreamed of their future, and the older ladies—" She put her hand on Thomas's shoulder and said, "They were taken back to the passion of their youth."

Jeb finally broke his silence. "Was it hard doing the dance?"

"Not knowing was harder than anything," Matt said.

"Were you not fearful?" Jeb asked.

"Not once I got out there," Matt said. "I knew what Jonathan told me about Grace."

"What did he say?" Grace asked, looking at her brother suspiciously.

"Jonathan said you danced like a princess," Matt said. "No lie." Jonathan nodded confidently. "And Will said that the man should make everyone focus on the lady."

"Good advice," Mary said. "Most watch the lady more than the man."

"I don't want to analyze it further," Matt declared. "My partner was a beautiful dancer. I don't think anyone was looking at me at all." Matt thought aloud. "Well, except for Henry Duncan. He's scheduled me for lessons when I return."

"You're going back into town?" Jeb asked.

"No," Matt said. "When I visit again from Philadelphia."

"I forgot that you're leaving soon," Jeb replied.

"When are you leaving, Mr. Miller?" asked Jonathan.

"It's the same," Matt said. "I need to wait for word about my ring."

"Is it a lot of money?" Jonathan asked.

"Not a polite question," his father said.

"I want to be sure Mr. Miller has enough to buy Thunder is all," Jonathan replied. "Thunder would be sad to be purchased by another." Matt smiled at the boy. He always asked more than simple questions.

"Jacob has offered to buy the ring," Matt said. "I'll have enough money to buy Thunder either way."

"I have asked Mr. Miller to help with the new fence," Thomas added. "He'll stay for a few more weeks after harvest."

"And you must finish your riding lessons," Grace added.

"Can't you do your apothecary in Richmond, Mr. Miller?" Jonathan asked.

"Richmond already has an apothecary, and there's not enough business for two," Matt said. "Besides, Philadelphia's my home."

They pulled into the gate. David and Faith had already returned. "Start soon on chores," Thomas reminded everyone. There was a collective sigh.

"Mr. Miller," Thomas said, "would you bring the horses in this evening?"

Matt nodded as the wagon stopped in front of the house. He hopped off after Grace and headed to the barn to change into his hiking clothes. Despite them now looking rather ugly, they were rugged and comfortable. He saw David as he was walking out. The man exaggerated his face into a frown and shook his head.

"What?" Matt exclaimed.

"You can minuet, too?" David said.

"Henry Duncan showed me," Matt replied. "I know what you'll say."

"What's that?" David was smiling.

"I'm a villain," Matt replied, "and my pulling Grace into that square with my knees knocking and my heart pounding was another part of my devious plan."

"Nah," David said. "I saw you look to the sky and pray as any common man. You may truly love my niece."

"I do," Matt said, "and I'll do the minuet in front of all of Virginia if I have to."

"Well done," David said. He reached out and shook Matt's hand.

The rest of the night was routine for Matt. There was comfort in joining the Taylors as they put the farm to sleep, which turned to foreboding as the night got closer to ending and he knew he would return to the barn. He had made his decision to stay in 1762, and tonight was the night he'd commit to his plan to end the rescue efforts from the future. He turned the phone on and the text came like clockwork.

"Are you there?" it read.

Matt typed, "I'm staying. Rescue one of the others."

"The portal will open at your original entry point in exactly 48 hours. Ordered to be there."

"I told you. I'm not coming."

"We can't risk you changing the timeline."

"Rescue one of the others. Again, the timeline might require my being in 1762."

"Be at the entry site in 48 hours. Step into the portal."

Matt typed, "Will not be there."

There was no reply.

# CHAPTER 41.

# TIME TO GO

---

It was Friday and Grace had finally given up on teaching Matt the finer points of jumping. After the Thursday lesson, he could get Thunder to jump over short obstacles, and that was all both man and horse were interested in doing. Thunder was so big that the earth and his body shuddered when he landed. Silver Star was a natural jumper, only two thirds the size of Thunder, with muscles wound tight around his bones like rubber bands. He looked like a pogo stick as he jumped, whereas Thunder took off and landed like a pallet of bricks.

Grace recognized their aversion to jumping in the ring early on Friday and decided that they should take their lesson into the country to learn to jump over natural obstacles. Matt was chasing after her as they headed toward the river. She was galloping, but not at full speed and he had no trouble keeping pace. She'd veer off the trail intermittently and hop her horse over logs or gullies. Thunder rarely hesitated on these jumps, which were inconsequential in comparison to the hurdles in the ring.

His long stride made it so he barely had to run to keep up with Silver Star.

In general, the week had been a peaceful one for Matt. Harvest was over, so the work was less frantic. He had spent most of the time hauling and cutting wood to build a fence for a new pasture. He and Charles had completed most of the task, with Jeb and Jonathan helping when their other chores allowed. Both boys had another week before they would return to school. Matt hadn't heard anything about his gold ring, but in truth, he hadn't given it much thought. He had little desire to dwell on the cascade of events that would occur once it sold. For the moment, he was happy to work, ride, and dream big dreams in his mini oasis.

Grace slowed Silver Star as they neared the mild precipice that looked down over the James River Valley. They hadn't been there since the day they first kissed. "Do you remember this place?" she said.

"How could I forget?" Matt said, smiling. "It was the first time you realized how charming I am."

"I may have realized before that," she said.

"You didn't act like it."

"Ladies need to think on such matters," Grace said. "You do realize that you're not the only eligible gentleman in Richmond?"

"With all these wealthy men chasing you, why waste your kiss on me?"

"Does it matter?"

"If I'm going to spend the next few years of my life working to win your hand, it does."

"The other men's lives are set," she said. "My life would have been decided."

He gave her a good-humored laugh. "So you like me because I have no plan."

"It's more than the fact that your future is uncertain," she said, frowning. "You desire to build a new life rather than live an old one."

"More true than you imagine," Matt said. "I think you also want to ride your horses."

"I'm not a petulant, spoiled child. 'Tis not all about me and my animals."

"I never said that!"

"I feel my husband must consider my wishes," she said. "Time will tell, I imagine." She thought for a moment. "I struggle much with the names you favor for our children."

"We can talk about it. You name one, I name one, that sort of thing."

"How many children did you expect to have?"

"Ten or fifteen, give or take."

He was joking, but Grace took him at his word. "Ten or fifteen? You must pledge to provide me a wonderful mansion with attached stables for me to agree to fifteen children."

"It's a challenge, then," Matt said. "Remember, I don't plan on providing you with anything. Both of us will build the mansion with attached stables. You get short breaks to have the kids and then it's back to work."

"Very charming," she said, frowning. She turned away and looked out over the river. "Where shall we live? I've been afraid to ask."

"Somewhere nearby. Close to your family, but not too close."

"Not too close?"

"Close enough to visit, but not so close that they are always involved in our decisions."

"A worthy conceit," Grace said. "A lady should command her own house. Mother should be near when I'm with child. I have some trepidation and will need her guidance. What if we only have daughters, like the Martins?"

"At least then you'd have to agree to a few of my names."

They led the horses closer to the water and wandered along, sharing dreams of the future and discussing even the most trivial items down to the smallest detail. At one point, they dismounted their horses and stood together overlooking the river and holding hands. Some unknown signal would make them turn to one another and kiss and then, as mysteriously, make them start walking again. The horses glanced at them occasionally while they grazed together on the long fall grass.

"We should get back," she said while they kissed.

"You're safe with me," Matt said. "I don't plan to ravish you."

"It's not you that worries me," she replied. He kissed her again and she pressed her body against him harder than she had before.

He was breathing her in.

She pulled away with passion in her eyes. "I'd give you everything now if you asked."

"You're not the only one who's vulnerable," Matt replied. "You better wait for me."

\*\*\*\*\*\*\*\*\*

They met Will in the road on their way home, so slowed their horses to walk beside him. He was returning from Richmond. "How was the lesson?" he asked.

"Very good," Grace replied. "How was your first week back in town?"

"Difficult," Will replied. "There were stacks of papers taller than me. I did a month's labor already."

"Did you visit Graine?" Grace asked.

"Wednesday for supper," Will replied. "Her sisters demanded to know my intentions."

"An alliance between our families would be beneficial," Grace replied.

"I admire her more and more," Will said. "Ofttimes I'll mention my work at Samuel's and she'll have some suggestion."

Matt looked back at him knowingly.

"I admit you said as much," Will acknowledged.

They were entering the gate to the farm. "Don't be late for supper," Grace called back to them. They watched her ride away, leaving them alone to walk their horses to the stable.

"You still off to Philadelphia?" Will asked.

"There's no way for me to start a business in Richmond."

"Have you spoken of your future with my sister?"

"Some," Matt replied. "I haven't heard from Jacob, so I have time."

"I imagined these would be good tidings," Will said tentatively. "Jacob said your ring has sold."

"Did he say how much?"

"No, but you owe me a meal."

"It sold for at least eighty pounds," Matt said. "I'll have enough money to pay for Thunder."

"When will you leave?"

"I want to attend church again, and then I'm off to Philadelphia."

"Jacob knows people who take the mail north," Will said. "Make sure you arrange an escort."

"How long is the trip to Philadelphia?"

"Ten days, perhaps," Will said. "You'll miss the farm."

"Nah," Matt said. "I'll meet friends in Philadelphia." Matt hoped the dread he felt didn't show in his voice. "Hopefully your sister will wait for me to come back."

"My sister's constitution has improved since you arrived," Will said, "and her smile has returned."

"Half the men in Richmond are trying to marry her."

"They've been trying to marry her since she was sixteen. Will rolled his eyes and laughed. "She's been waiting for someone who is not vexed by a lady wearing breeches."

"I think if I were your father, I'd find every way to steer her elsewhere."

"We all have high expectations for the man that will marry Grace. Is she worth your effort?"

"Effort?"

"Think on it," Will replied. "Tomorrow you can ride away without encumbrance, never having to impress Father again."

"I'd regret riding away from your sister."

"Then I'd welcome you as my brother," Will replied. He looked down at Scout. "Can I borrow the dog for a bone?"

"He's your dog," Matt said.

"Should you return for my sister's hand, I suspect he should desire to live with the Millers, especially if you take his horse."

"I hadn't thought about that," Matt said. It was true. Scout was going to miss Thunder.

"Come, boy," Will said. "Let's get a bone."

Matt was embarrassed when the dog looked up at him for permission. "Go, dog," he said, and Scout followed Will to the house.

Matt continued on to the corral. He unsaddled Thunder, brushed him briefly, and let him into the pasture. He had a million thoughts as he washed up and put on a new shirt for dinner. He sat on his bed, looked around the barn and then up at the ceiling. The first time he woke up and stared at that ceiling seemed like a lifetime ago. He gazed up into the empty space, hoping that he was part of some plan. *Give me strength.*

**********

Dinner that night was lively. "Are you going back into town tomorrow, Mr. Miller?" Thomas asked.

"My ring sold," Matt replied.

"'Tis good news," Thomas declared. He looked at his daughter to gauge her reaction.

"I'll have money to pay for Thunder and fund my trip," Matt said. "I'll be traveling back to Philadelphia soon."

"This is a surprise," Grace said. "When will you leave?"

"Monday. I'd like to go to church one more time."

"Find a church in Philadelphia," Thomas said.

"It's important for a young man to attend church," Matt replied. "I was listening."

"Children want church," Mary added. "Their appreciation for God's gifts can only be learned from a father who is influential in this regard."

"I have not always gone to church and been a godly man," Thomas admitted. "A man's success and joy in life begins and ends with a life centered on the Lord. If you move towards Him, He moves towards you."

"You've convinced me," Matt said.

"Motivations matter little," Thomas said. "Put yourself in the Lord's house and your blessings will come."

# CHAPTER 42.

## TOOTHPASTE, PART V

---

"We don't know what became of the others," Kevin Moore said. "What if we get him back and he's dead. Then what?"

"He made it through the first time," Brian Palmer replied. "You either believe the data or you don't."

"He said that he was sick," Jacob Cromwell said. "It might be worse coming back."

"Data!" Palmer exclaimed. "He's fine."

David Greer, who had been listening quietly, felt the need to speak. "Brian, you keep saying 'data,' but I don't think we have any. We may have been responsible for the deaths of three others."

"That's nonsense," Palmer replied. "These people are alive."

"That's your theory," Greer said. "I can't imagine how anyone survives an unprotected trip through a wormhole."

"We have contact with one," Palmer said. "That's proof enough."

"We should shut this down," Cromwell said. "It was an accident; I can still live with myself. The consequences of anything we do in the future are on us."

"I'm sick of working nights and never getting any credit," Greer said. "I'm done."

"Me too," Moore added. "From the things Colonel Gabriel has said lately, he'll support this."

"You're reading into his comments," Palmer replied.

"I'll pull together a meeting tomorrow," Cromwell said. "I don't want to work on this anymore. Besides, Matthew Miller says he doesn't want to be rescued."

"Fine," Palmer replied, resigned. "We can talk to Colonel Gabriel in the morning."

"Let's go have a beer," Moore said. "I don't want to end this on a miserable note. We're a great team...we just need a project we can talk about."

"Fine," Palmer agreed. "I'm getting tired of being the only one who wants this."

"We can make plans at the bar," Cromwell said. "I have a few opinions about what we should work on next."

"Faster than light is impossible," Moore said.

"Give me a chance to explain over a few beers!" Cromwell replied.

"I haven't submitted my abstract to the APC yet," Palmer said. "I'll meet you guys in a half hour or so."

The four physicists did a quick cleanup of the lab and switched off the instruments. As the lab grew quiet, a look of relief began to grow on their faces with the idea that they were making the right decision in shutting down the project. Palmer waved to them as they left. "See you in a moment. Shouldn't take too long."

Palmer sat there working on the computer until he heard the outside door click closed, and then he sprang into action. He walked systematically around the laboratory, flipping switches. The stainless steel reactor began its calm hum and the room filled with soft green light. He set the chronometer to 1762 and slowly released particles into the reflectors. Palmer sat at his computer and typed, "Are you there?"

"This will be my last communication," Matt replied.

Palmer typed, "I'll be opening up the portal at your original entry point. Please step in."

"I'm not near the entry point."

"When can you get there?"

"Tell my dad I love him and I'm okay. I'm staying."

"Not if I can help it," Palmer said aloud. The longitude and latitude of Matt's cell phone was flashing in the corner of the computer monitor. Palmer now had the exact position for the cell transmission. He filled in the values and flipped the switch. The particles shot into the reflectors at full power and the time portal on the wall began to glow.

**********

Matt Miller was in the barn, leaning against a stack of hay bales, watching the symbol on the screen of his phone as it shut down, and talking to the dog. "Hopefully, these guys have gotten the message. I'll take my chances in 1762."

The shock the phone gave him was a familiar one, and suddenly his memory of the events leading up to his disappearance from the hiking trail came rushing back. He knew immediately what was happening. "They're trying to open a hole on top of me," he said aloud. He tossed the phone to the corner of the barn, dashed toward the dog, picked him up and dived into the opposite corner. Scout yelped angrily as they hit the floor. Matt looked up in time to see a faint green field appear above the haystack where they had been standing. Three seconds later, there was a sharp crack like a bullwhip and then complete silence. About forty hay bales had disappeared and there was a large round hole in the barn roof. Hay from surrounding stacks tumbled to fill the void left by the missing bales.

"Sorry about that, dog," Matt said. "You okay?" The dog was already on his feet, sniffing the ozone in the air. There was knocking on the barn door and then it slid open. Matt was still sitting dazed on the floor.

David walked in and stared at the damage as he made his way to Matt and the dog. "What in God's name was that?" he exclaimed. "Are you hurt?"

"Lightning came through the roof and hit the bales," Matt lied. "I'm fine."

David looked up through the hole at the moon. "Lucky it didn't start the barn on fire."

"Yeah, I guess," Matt replied, still shaken.

"Everything looks fine otherwise," David said. "We can get this cleaned up tomorrow morning. You fine to sleep in here?"

"Probably. Lighting doesn't strike twice in the same place, right?"

"I imagine not," David said, looking back up at the roof. "There's not a cloud in the sky."

"Goodnight, David," Matt said quickly. He was hoping the man would not start asking him a bunch of probing questions. David took his cue, leaving Matt and Scout in the barn. Matt walked to retrieve the cell phone from the corner. It had a few new scratches. He checked to make sure it was off, then set it on the farthest bench and covered it with his folded clothes. He was mostly sure it had to be transmitting to be tracked, but he'd give it a few days before turning it on. He wanted it far away for tonight.

**********

Back in the twenty-first century, military police were prying open the door to the Oak Ridge Propulsion Laboratory. Brian Palmer was still trapped inside and making frantic pleas for help. A team of men in radioactive hazmat suits was waiting in the parking lot after being mobilized by NORAD, which had registered a nuclear explosion coming from the site. Upon arrival, they realized that there was no evidence of an explosion. They scanned the area and verified that it was clean and were in the process of packing up when the MPs started to arrive. Many of the hazmat men sat waiting for an explanation of the false alarm.

Colonel Gabriel called the three physicists and interrupted their trip to the bar. They drove to the lab immediately and stood there as MPs inspected the laboratory door. "Anything dangerous in there?" one MP asked.

"There's a reactor," Jacob Cromwell replied. "It doesn't sound like it's running, so probably no."

The other MP, who had been working on the door, said, "Got it open." It was wide enough for someone to squeeze through.

The MP gestured to Cromwell. "You first. You're the expert."

"Fine," Cromwell replied. All he could think of was his infant son and how he'd failed him. He stooped under a shelving unit that had fallen against the door and squeezed through. Bales of hay and pieces of weathered wood filled the laboratory. Cromwell paused, stunned.

"Don't stand there," Palmer said. "Get this off me." He was trapped facedown under a small mountain of hay bales. The MPs were trying to squeeze into the room.

"Where's all this hay from?" Cromwell asked.

"1762," Palmer said. "Where do you think?"

"You transported all this hay back from 1762?" Cromwell replied. "What for?"

"It wasn't on purpose," Palmer said. "I tried to pull Matthew Miller home."

Cromwell couldn't believe how angry he was. "You're a damn fool."

"I know," Palmer replied. "He asked us to tell his dad he's okay."

# CHAPTER 43.

## NOW WHAT?

David was quiet for the beginning of their journey into town, but about halfway there, thoughts came pouring out of his mouth. "You sleep well after all that commotion in the barn," he asked.

"Like a baby," Matt replied. It was a lie. He had kept one eye open all night waiting for another green beam.

"Fortunately you didn't get hit."

"I can't help it if I'm lucky," Matt said. He gave David a big smile to emphasize the jest.

"You think you'll have enough money to board your horse in Philadelphia?"

"I should have an extra ten pounds. I'll use that."

"You'll ask Jacob about traveling with someone? You don't want to travel alone."

"I'll arrange it."

"That girl would be unbearable should you get scalped."

"I'll try not to get scalped," Matt said.

"I have trouble saying things like this."

"I believe you," Matt said, smiling.

"All men deserve a chance. You have yours."

"Still feels like the odds are against me," Matt said.

"Dreams of young men are always thus," David replied.

**\*\*\*\*\*\*\*\*\***

They arrived in Richmond soon afterwards. "I'm off to Jacob's," Matt said. He arranged to meet David after lunch for their journey home. Matt would go first to the silversmith's, then to a saddler with the hopes of buying a saddlebag for his journey, and then visit Henry and buy him lunch.

Jacob Berkley greeted Matt with a hearty hello. "It sold," he said with a wide smile on his face.

Matt looked around to make sure the store was empty and that they spoke in private. "How much?"

"Better sit," Berkley said, laughing. "You should be prepared for some disappointment."

"It sold," Matt replied. "The possibility for me to be disappointed is over."

"Seven hundred pounds."

"Seven hundred pounds?" Matt exclaimed. "You're joking."

"No. I have never been more wrong about a piece of jewelry. They took eight percent for commission and secure courier."

"Seven hundred pounds?" Matt repeated. He was sure it was a joke.

"I have the bill of sale," Berkley said. "Three hundred twenty pounds each after expenses."

"Three hundred and twenty pounds?" Matt said, dumbfounded. "Who bought it, the king of England?"

"One of his representatives," Berkley said.

"You're joking," Matt repeated. He was still waiting for the punch line.

"It was the crown carved in the stone they desired. Have you a purse for the money?"

"I'm buying a bag. I'll pick it up on my way out of town."

"I take trade all day," Berkley replied. "I have a stack of Joes."

"Joes?"

"Six pounds each."

"I'll be back," Matt said. "I'm traveling to Philadelphia and need an escort. Thomas Taylor mentioned you know people."

"Two brothers named Ezekiel and Robert Wilkins go north with the post on Monday. You'll receive safe passage for a modest fee."

"I wish to join them this Monday if you wouldn't mind arranging it," Matt said. He reached his hand out. "Good doing business with you."

\*\*\*\*\*\*\*\*\*\*

Matt returned for his money after completing his errands and buying a new saddlebag.

"All ready to go?" David said as Matt walked out of the silversmith's store.

"Let me pack this gold away," Matt said. "So much depends on it."

David waited for Matt to drop the leather pouch into the saddlebag and mount Thunder, and then they started the journey home. They rode in silence for a long while.

"You know, naught depends on all that gold," David finally said.

Matt laughed. "It's the missing piece of the puzzle," he proclaimed.

"How so?"

"If I were wealthy, I'd already be courting Grace."

"You don't believe that."

"You can't deny I'd be better off if I had money."

"You would be better off if you had *earned* money."

Matt could hear the irritation in David's voice. "What's gotten into you?"

"In two years, you should be able to give all that money to the poor and have no doubt that you can make it all back. That's the man Thomas is seeking for his daughter." He sighed and smiled. "I'm going on and on. You're young, and I'm sorry if I let you vex me."

"Apology accepted, I guess," Matt said. "I never believed it was only about the money."

"Seems the ring sold at a good price," David said.

"More than expected," Matt replied. "There's no excuse for me to stay."

"Afraid to go home?"

"There's nothing waiting for me in Philadelphia. It's not going to be easy."

"When has it ever been easy to walk the path of the Lord?"

They were quiet for the rest of the trip. The only sounds were those of the horses and the jingle from the coins in Matt's saddlebag. Matt made a mental note to secure them for his journey to Philadelphia so they wouldn't attract attention.

\*\*\*\*\*\*\*\*\*

The remainder of the day moved in fast-forward. Before Matt knew it, it was the next morning and they readied themselves for church. He sat next to Grace during the service, and this moment too seemed to disappear in an instant. After church, he was able to convince the Taylors to let him buy them lunch. He said his final goodbye to Will, who stayed in town. Matt rode back with the Taylors, arriving in time for one last ride in the country with Grace. Afterwards, he found himself sitting on the porch, petting and talking softly to the dog until it was time for dinner. Then it was dark and time for bed. He talked briefly to Faith and David, thanking them and saying goodbye.

"I desire to walk Mr. Miller to the barn and say goodnight," Grace said to her father. Thomas looked at them, considered it for a moment, and said a simple "Fine."

As they opened the door, Mary called out, "Jonathan, would you please accompany Mr. Miller and Grace to the barn? I don't want Grace to be lonely when she walks back by herself." The boy nodded and followed them out the door.

"I must watch for improprieties again," Jonathan said. "My part is more important now."

"Why's that?" Matt said.

"Grace doesn't pretend not to admire you."

Grace jabbed him in the stomach. "You shouldn't talk about improprieties in front of a lady," she said. They broke into laughter and poked each other as they walked. When they reached the barn door, all three stood there awkwardly.

The boy finally looked up and said, "I saw fireflies on the other side of the barn. I'll walk there and spend time catching them. You never know when I shall return." He gave Matt an exaggerated wink and then walked around the barn.

"I'll miss that boy," Matt said.

"And he you," Grace replied. She leaned into him and kissed him softly and he pulled her close.

"I'll do my best to write every day," he said. "If you don't hear from me for a few days, though, don't assume it's fine to marry some spoiled rich boy."

"Don't go too many days without a letter, then," she replied.

It seemed only an instant before Jonathan had returned, but Matt knew it had been a long time. The thought occurred to him that every moment he spent in the presence of this woman he loved was definitive proof of Einstein's theory regarding the relativity of time. "I was only able to catch a few," Jonathan called as he rounded the corner. "Were you two able to catch any?" Grace poked him repeatedly when he stepped into range. Both giggled again.

Matt reached out and shook the boy's hand. "I guess this is goodbye, my man," he said. "I'll be leaving early tomorrow."

"You'll be back soon, though?" Jonathan said.

"As fast as I can."

Jonathan glanced over at his sister and grabbed her hand. "Mr. Miller will be back." He pulled her toward the house and she waved as he dragged her away.

"Thanks again, dude," Matt called out to the boy.

"It's Jonathan."

"I know."

"I was the one who found him," the boy said as they faded away into the dark.

********

Matt sat for a long time in the barn, petting and talking to the dog. He wasn't sure what time it was when he finally went to sleep, but like everything else in those last days, morning came too quickly. He washed, ate breakfast with the Taylors, and went out and saddled Thunder while Scout looked on. Most of his things fit in his saddlebag, so he left his backpack tucked away in the barn,